INFINITY IN THE PALM OF YOUR HAND

INFINITY IN THE PALM OF YOUR HAND

*Fifty Wonders That Reveal an
Extraordinary Universe*

MARCUS CHOWN

DIVERSION
BOOKS

Diversion Books

A Division of Diversion Publishing Corp.

443 Park Avenue South, Suite 1004

New York, New York 10016

www.DiversionBooks.com

For more information, email info@diversionbooks.com

Book design by Pauline Neuwirth, Neuwirth & Associates.

First Diversion Books edition April 2019.

Paperback ISBN: 978-1-63576-594-6

eBook ISBN: 978-1-63576-593-9

First published in the United Kingdom by Michael O'Mara Books.

Printed in the U.S.A.

1 3 5 7 9 10 8 6 4 2

To Allison, Colin, Rosie, Tim, and Ornella
With love, Marcus

CONTENTS

CONTENTS

PART FOUR: SOLAR SYSTEM THINGS

PART FIVE: FUNDAMENTAL THINGS

PART SIX: EXTRATERRESTIAL THINGS

CONTENTS

FOREWORD

"Nothing is too wonderful to be true."
—MICHAEL FARADAY

COMEDIANS, WHEN INTRODUCED AT parties as comedians, may feel under pressure to tell a joke. Science writers, when introduced at parties as science writers, may feel under pressure to trot out a jaw-dropping scientific fact. Well, I do. Sometimes.

What kind of thing should I say? Something short and snappy. Enough to intrigue, make a person smile, but not enough to cause their eyes to glaze over so that I inadvertently appear a bore.

I try things out on my wife, who has no science background, often while she is watching TV: "Did you know that an electron rotated through three hundred and sixty degrees is not the same electron?"

"Um," she says, not turning away from the screen.

"What about: you could fit the entire human race in the volume of a sugar cube?"

"Yes, that'll do. Now, *can I watch my program*?"

My wife is an important sounding board.

There is another reason for finding these intriguing one-liners, though, and that's public talks.

Many talks I give are during tours to promote one of my books. The problem is that it is impossible to do justice to an entire book in forty-five minutes or so. Instead, therefore, I often pull out some intriguing facts and use them not only as a means to catch people's interest but also as a way into describing some of the science I have written about.

It all started with my book *What a Wonderful World: Life, the Universe and Everything in a Nutshell*, which was supposed to be about *everything*—though that is, of course, impossible. It did, however, cover everything from finance to thermodynamics, holography to human evolution, and sex to the search for extraterrestrial intelligence. What, I wondered, should I include in my talk, and what should I leave out? It was then that I got the idea of talking about my "Top 10 bonkers things about the world."

The great thing was that this was a movable feast. So, if the audience looked bored with one of my bonkers things, I would drop it from the next talk and include something else that would hopefully get a better reception. I imagine this is a bit like being a stand-up comedian. If a joke does not work one night, it gets discarded and substituted for something else for the next performance.

And the beauty of the format is that it works for other subjects as well. I developed an app called "Solar System for iPad," which was followed by a book called *Solar System*. In talks to promote it, I talked about my "Top 10 bonkers things about the solar system."

Which brings me, finally, to this book. Why not, I thought, put together some of the most mind-blowing scientific facts I have discovered over the years—things I have covered in books and articles and things I've never written about before—and use them as a way into explaining some thought-provoking and often deeply profound science?

For instance, the fact that, if you squeezed all the empty space out of all the people in the world, you could fit the human race in the volume of a sugar cube illustrates perfectly the mindboggling emptiness of matter. You, me, everyone—we are all pretty much ghosts. And that leads naturally on to quantum theory, the most successful but also the weirdest physical theory ever devised, which ultimately provides the explanation of why atoms are overwhelmingly made of nothingness. The fact that, if the sun were made of bananas, it would be precisely as hot as it is now leads to the remarkable fact that the temperature of the sun has nothing whatsoever to do with what is powering it. And the fact that 95 percent of the universe is invisible leads to, well, the extraordinary—in fact, embarrassing—realization that everything scientists have been studying these past 350 years amounts to no more than a minor constituent of the universe. And—even worse—we have pretty much no idea what the major component is.

Years ago, I interviewed the American planetary scientist and science popularizer Carl Sagan at the Dorchester hotel in London (his suite, I remember, had fantastic views of Hyde Park and the Serpentine). After writing nonfiction books like *The Cosmic Connection*, Sagan had written his first science-fiction novel, *Contact*, which would later become a film starring Jodie Foster. I asked him what he preferred: science or science fiction. Without the slightest hesitation, Sagan replied: "Science. Because science is stranger than science fiction." And it is. We find ourselves in a universe far stranger than anything we could possibly have invented. I hope that, in the following pages, I manage to convey some of this strangeness—and wonder.

I really enjoyed writing this book. And I hope you enjoy reading it. At the bare minimum, I hope it will arm you with a few amazing facts about the universe to trot out at parties.

MARCUS CHOWN, London, 2018

PART ONE

BIOLOGICAL THINGS

1.

THE COMMON THREAD

You are a third mushroom

"How extremely stupid not to have thought of that."
—THOMAS HUXLEY,
ON HEARING OF DARWIN'S THEORY OF
EVOLUTION BY NATURAL SELECTION

YOU ARE ONE THIRD mushroom. That's right. You, me, all of us share a third of our DNA with fungi (as if my Christmas-card list was not long enough already!). This is strong evidence that humans and mushrooms—in fact all creatures that share the earth today—have a common ancestor. The person who first recognized this was the English naturalist Charles Darwin.

In 1831, aged just twenty-two, Darwin took up the post of ship's naturalist on HMS *Beagle*. During its five-year voyage, he made a series of striking zoological observations. He noticed, for instance, that the birds and animals on the isolated Galápagos Islands, 1,000 kilometers off the west coast of South America, appeared to be variants of a small subset of birds and animals found on the continent. Not only that, but the birds and animals on each island of the Galápagos archipelago also differed from each other in subtle ways. Most famously, the finches that lived on islands where large nuts were available had stubbier beaks than finches on other islands.

After eighteen months of intense concentration, a light went on in Darwin's mind. He realized why creatures were so exquisitely tailored for their environments. And it was not, as was the prevailing view, that they had been "designed" by a Creator. There was a perfectly natural mechanism that created the "illusion of design."

Most creatures, Darwin recognized, produced many more offspring than could be supported by the available food and were therefore destined to starve to death. However, in the struggle for survival, those individuals best suited to exploit the resources of their environment persisted, whereas those least suited perished. The casualties were staggeringly huge. But, by this process of evolution by natural selection, creatures changed incrementally, generation by generation, to be better adapted to their environments.

Darwin reasoned that, millions of years before, when the volcanic Galápagos Islands had risen from the sea, a handful of creatures—birds that had flown and other animals that had been driven by storms across the ocean on mats of vegetation—had reached the archipelago from the mainland of South America. Finding an essentially empty world, they had spread out to fill all the available ecological niches. Darwin's finches, isolated on different islands, had suffered the pressure of natural selection; the least adapted for survival had been brutally culled while the best adapted had prevailed. In the case of an island with large nuts, inevitably the finches that survived were variants with tough stubby beaks, perfect for cracking open big nuts.

Darwin's courage was to present his theory of evolution by natural selection without knowing two key things: first, how characteristics were passed on, or inherited, from generation to generation; and, second, what created the variation in offspring—the raw material for natural selection to work on. We now know that these two things are intimately connected. The blueprint for an organism is recorded in the large biological molecule called deoxyribonucleic acid, or DNA, which is carried in every cell.[1,2] And it is mutations in DNA, often caused during the copying process, when cells reproduce, that give rise to varied and novel traits in offspring. "The capacity to blunder slightly is the real marvel of DNA," said the American biologist Lewis Thomas. "Without this special attribute, we would still be anaerobic bacteria and there would be no music."

According to Darwin, all creatures on Earth today have evolved by a process of natural selection from a simple common ancestral organism. This, ultimately, is the reason why we share one third of our DNA with mushrooms. In fact, the following stretch of DNA is present in every cell of every creature on Earth, including every one of the one hundred trillion cells in your body: GTGCCAGCAGC-CGCGGTAATTCCAGCT CCAATAGCGTATATTAAAGTTGCT-GCAGTTAAAAAG.[3] Can there be a more striking piece of evidence that all creatures are related and that they evolved from a common ancestor, exactly as Darwin claimed? In the words of Thomas: "All of today's DNA, strung through all the cells of the earth, is simply an extension and elaboration of the first molecule."[4]

Darwin knew that the process of evolution by natural selection was painfully slow and would have required hundreds of millions, if not billions of years to create the profusion of life on Earth today. The first tentative evidence of life on our planet dates to about 3.8 billion years ago. Conceivably, the first cell—dubbed the "last universal common ancestor," or LUCA—arose around four billion years ago, a mere half a billion years after the birth of the earth. Exactly how this happened—and how the step from nonlife to life was taken—remains one of the biggest unanswered questions in science.

2.

CATCH ME IF YOU CAN

Some slime molds have thirteen sexes

"I admit, I have a tremendous sex drive.
My boyfriend lives forty miles away."
—PHYLLIS DILLER

SOME SLIME MOLDS HAVE thirteen sexes. (And you think you have trouble finding and keeping a partner!) Their sex cells, unlike human sperms and ova, which are hugely different in size, come in only one size. The gender of the cells is instead determined by three genes known as MatA, MatB, and MatC, which come in a number of variants. In fact, there are so many variants that potentially it is possible to have more than five hundred different sexes. To reproduce, a slime-mold spore must simply find a partner with different variants of its three genes.[1]

Nobody knows why some slime molds have thirteen sexes and some five hundred-plus. But then nobody knows why we have two sexes. Nor, for that matter, why we have sex.

In evolutionary terms the name of the game is to get your genes into the next generation.[2] Not some of your genes but all of them. The sensible thing would therefore be to clone yourself since this ensures the transference of 100 percent of your genes to any offspring. Such asexual reproduction is in fact what most creatures on Earth practice. Organisms that have sex, on the other hand, pass on only 50 percent of their genes to the next generation. This means not only that they must give birth to twice as many offspring to achieve the same as asexual organisms but they must expend extra energy finding a partner as well. Sex appears to make no sense at all.

Many explanations for sex have been proposed but, until recently, none has been convincing. One, however, has now gained increasing acceptance—and, surprisingly it concerns parasites.

Across the world at any one time, more than two billion people are unfortunately infected with parasites, which range from intestinal worms to malarial parasites. Such parasites tend to be small and able to reproduce quickly, which means they can go through many generations during the lifetime of their host. As a consequence, they can quickly adapt to their host so that they efficiently exploit its resources. The exploitation of those resources, however, is at the expense of their host, which is not only weakened but sometimes even killed.

Understanding what sex has got to do with parasites takes a bit of background. Imagine the DNA of an organism to be like a deck of cards. When the organism clones itself, its offspring inherit the entire deck of cards with maybe one or two cards slightly changed due to a random mutation. By contrast, in sexual reproduction, offspring inherit half a deck of cards from one parent shuffled together with half a deck of cards from the other parent. This makes the offspring not only different from either parent but also utterly unique. Consequently, the parents' parasites find themselves ill-adapted to the offspring and die.

The idea that sex continually wrong-foots parasites was proposed by the American biologist Leigh Van Valen in 1973.[3] Basically the idea is that, although parasites can change rapidly, a host population can survive their relentless onslaught by changing even more rapidly.

In *Through the Looking Glass*, Lewis Carroll's 1871 sequel to *Alice in Wonderland*, Alice runs beside the Red Queen but cannot understand why she is making no progress.

> "In our country," said Alice, still panting a little, "you'd generally get to somewhere else—if you run very fast for a long time, as we've been doing."
>
> "A slow sort of country!" said the Queen. "Now, here, you see, it takes all the running you can do, to keep in the same place."

Van Valen's parasite explanation for sex, which has become known as the "Red Queen Hypothesis," received strong observational support in 2011.[4] By genetic manipulation, biologists in the United States engineered two different populations of the roundworm *Caenorhabditis elegans* to reproduce in two different ways: one asexually, by fertilizing its own eggs; the other sexually, by female and male worms mating.[5] The biologists then infected both groups of worms with a pathogenic bacteria. *Serratia marcescens* rapidly drove extinct the self-fertilizing population of worms but not the sexually reproducing ones. These outpaced their coevolving parasites—they continually ran faster. So, although it may not be the most romantic explanation for falling in love, defense against parasites seems to be the reason for sex.

3.

THE OXYGEN TRICK

Babies are powered by rocket fuel

"In every one of us there is a living process of combustion
going on very similar to that of a candle."
—MICHAEL FARADAY[1]

A BABY SQUIRMS IN A cot. A rocket climbs high in the sky on a column of smoke and flame. Not much connection between them, you might think. But you would be wrong. Both are energized by the same chemical reaction. Both are powered by rocket fuel.

This is actually not as surprising as it may appear. Boosting a heavy rocket into orbit requires the most powerful fuel—the one that, pound for pound, packs the biggest oomph. Life on Earth has been engaged in almost four billion years of trial-and-error experimentation. It would be odd if, in the attempt to power biological processes, it had not stumbled on the most potent energy source possible.

That energy source is the chemical reaction between hydrogen and oxygen—or *combustion*, as it is more commonly known. In the case of all animals, the hydrogen is extracted from food and the oxygen from the air. In the case of a rocket, liquid hydrogen and liquid oxygen are supplied by humans. Understanding how the reaction between hydrogen and oxygen works and where the tremendous energy comes from requires a little background.

Atoms of hydrogen and oxygen—in fact, all atoms—consist of a tiny nucleus and even tinier electrons. The electrons orbit the nucleus, snared by its powerful electric force, in much the same way that planets orbit the sun, gripped by its gravity.

There is a general tendency in physics for things to minimize their *potential energy*, which is equivalent to doing something useful with it—or, in the scientific jargon, doing "work." For instance, a ball high on a hillside has high gravitational potential energy. Given the opportunity, it will try to minimize it by rolling to the bottom of the hill where it has low gravitational potential energy. Similarly, the electrons in an atom, as surely as balls rolling down a hill, will try to minimize their energy.

When two atoms come together, there are new ways for the electrons in each atom to arrange themselves. If a configuration exists with a lower total energy than in the two atoms separately, then the atoms will combine to form a molecule. In a nutshell, this is all chemistry is: the rearrangement of electrons.

Since the energy of the molecule is less than the energy of the separate atoms, energy is left over. It is a cornerstone of physics that energy can be neither created nor destroyed but only transformed from one form to another—for instance, from electrical energy into light energy. Consequently, the surplus energy is available to do things.

In a rocket, the reaction between a hydrogen atom and an oxygen atom—specifically, two hydrogen atoms reacting with each oxygen atom to make H_2O (water)—liberates a large amount of energy. This heats the water and expels the white vapour at great speed from the back of the rocket. (Ultimately, rockets are steam-powered!) Action and reaction being equal and opposite, the high-speed exhaust propels the rocket forward.

The liberation of so much energy by the reaction between hydrogen and oxygen is the reason a rocket can be lifted to the edge of space.[2] It is also the reason a baby—not to mention you and me and every last animal on Earth—can do all the things it can do.

In a rocket, hydrogen and oxygen combine to make water with the explosive release of a large amount of heat energy. Living organisms clearly do not use such a violent process. Instead, they liberate the energy, step-by-step, in a far more subtle and less destructive way.

What happens when hydrogen and oxygen react together in a rocket is what happens in all chemical reactions—the electrons play a game of musical chairs. Specifically, an oxygen atom grabs electrons from two hydrogen atoms.[3] In the process, the oxygen and hydrogen atoms fuse into a molecule of water. However, in biology's twist on the oxygen-hydrogen reaction, the hydrogen atoms in a cell supply electrons to an oxygen atom, but the two atoms never actually meet. Between the hydrogen atoms and the oxygen atom stretches a long wire made of protein complexes. And the donated electrons, bursting with excess energy, hop from location to location down the wire.

As each electron hops down the wire, it drives hydrogen nuclei, or *protons*, through channels, or pores, in the cell membrane.[4,5] Since protons carry a positive electric charge—the opposite of electrons—this charges up one side of the membrane with respect to the other. Something similar happens in a battery, creating an electric force field between its terminals. And this hints at what the super-energetic electron does as it thunders down the protein wire to an oxygen atom: it turns the cell membrane into a charged-up battery. The resulting electric force field across the membrane is extremely power-ful—comparable to the field in a thunderstorm that breaks down the atoms in the air and unleashes a bolt of lightning.

Fortunately, the cells in your body do not crackle with lightning. This is because the tremendous electric force field extends over only the tiny thickness of a cell membrane—about five millionths of a millimeter—and other molecules intervene to stop this force field having its way.

The powerful electric force field of the membrane battery drives a chemical reaction that creates adenosine triphosphate, or ATP. Such molecules are stores of energy—portable batteries, if you like. So, as the electron bounces down the protein wire, losing energy all the while, it leaves in its wake a large number of energy-packed ATP molecules. Released into the wild, these can power cellular processes wherever and whenever necessary.

In the final analysis, you are battery powered! There are about a billion ATP molecules in your body, and they are all used and recy-cled every one to two minutes. Toys may need a handful of batteries, which are used up in a few hours. By contrast, your body uses up ten million power packs every second.

Eventually, the electron arrives at the end of the protein wire, sapped of its energy. There it combines with the waiting oxygen atom. And, when a second electron from another hydrogen atom joins it, the oxygen atom achieves the highly desirable state of a filled outer shell of electrons. But this is not quite the end of the story.

As mentioned earlier, living things obtain their hydrogen from food. An amazingly subtle and energy-efficient process inside a cell, called the *Krebs cycle*, strips hydrogen atoms from food—that is, from either molecules of sugar (glucose—$C_6H_{12}O_6$) or fat. This pro-cess leaves behind carbon atoms. And if the oxygen atoms, now with their filled shell of electrons, share those electrons with a carbon atom, the result is a very stable molecule of carbon dioxide (CO_2). Carbon dioxide, along with water vapour, is what oxygen-breathing animals breathe out as waste.

So, there you have it. Your body takes hydrogen from your food and robs its fizzing electrons of every last drop of energy before donating them, utterly exhausted, to oxygen. In a nutshell, that is the power source of babies—and of all life.

4.

SEVEN-YEAR ITCH

Today your body will build about
300 billion cells

"Not a single one of the cells that compose you
knows who you are, or cares."
—DANIEL DENNETT[1]

TODAY YOUR BODY WILL build about 300 billion cells. That is
more cells than there are stars in our Milky Way galaxy. No wonder I feel exhausted doing nothing!

A cell is a tiny transparent bag of gloop. It is the atom of biology.
In fact, it is fair to say that there is no life except cellular life. The first
fossil evidence of cells dates from about 3.5 billion years ago and the
first evidence of chemical changes caused by life from about 3.8 billion years ago, making the likely origin of life sometime around four
billion years ago, a mere half a billion years after the birth of the earth.

Every human being is a giant colony of cells. "A good case can be
made for our non-existence as entities," said the American biologist
Lewis Thomas.[2] "We are, each of us, a multitude," said Carl Sagan.[3]
That multitude is in fact about 100 million million cells—a truly
astronomical number. I am a galaxy. You are a galaxy. In fact, each
of us is a thousand galaxies since there are more cells in your body
than there are stars in a thousand Milky Ways.

Each of the cells that make up a human being is a miniature world as complex as a major city. It seethes with the activity of billions of nanomachines. It boasts administrative centers, workshops, storehouses, and streets clogged with ceaseless traffic. "Power plants generate the cell's energy," says American journalist Peter Gwynne. "Factories produce proteins, vital units of chemical commerce. Complex transportation systems guide specific chemicals from point to point within the cell and beyond. Sentries at the barricades control the export and import markets, and monitor the outside world for signs of danger. Disciplined biological armies stand ready to grapple with invaders. A centralized genetic government maintains order."[4]

Our lives begin, however, as a single cell when a sperm, the smallest cell in the body, fuses with an ovum, the biggest cell in the body. Every human, in fact, spends about half an hour as a single cell. (I remember it being very dull—I couldn't wait to find another cell to play with!) The single cell splits into two. This is a phenomenal process. In only half an hour, not only does a cell make a copy of its DNA—a process that, for speed, occurs simultaneously at multiple sites on the DNA—but it also constructs something like ten billion complex proteins.[5] This is more than 100,000 a second. In an hour, the two cells split into four, then later eight, and so on. After several divisions, chemical differences across the developing embryo cause the cells to differentiate. The process culminates in cells "knowing" they have to be liver cells or brain cells or bone cells. Eventually, the single initial cell proliferates into the 76 million million cells of the human body.

But this is not the end of the story. Apart from brain cells, few of the cells in your body are permanent. For instance, the cells that line the wall of your stomach are bathed in hydrochloric acid, which is strong enough to dissolve a razor blade, so they must be remade constantly. You get a new stomach lining every three or four days. Blood cells last longer but even they self-destruct after about four months. In fact, almost all the cells in your body get replaced every seven years. Perhaps this explains the famous "seven-year itch." You look at your partner and think, "Hey, that's not the person I got together with."

5.

LIVING WITH THE ALIEN

You are born 100 percent human
but die 50 percent alien

"There's someone in my head and it's not me."
—PINK FLOYD

HALF THE CELLS IN your body do not belong to you. The fraction was previously thought to be as high as nine-tenths, but a recent study scaled it down.[1] Nevertheless, the realization that 50 percent of the cells in your body are not yours—around 38 million million of them—is a remarkable thing.

Some of the foreign cells hitching a ride on you are fungi; some are bacteria. In fact, you would not be able to extract nutrients from your food if it were not for hundreds of species of bacteria living in your stomach. This is why you can end up with diarrhea if you take antibiotics. Being indiscriminate in their action, such drugs kill not only disease-causing bacteria but bacteria that you need as well.

Bacteria are much smaller than the cells in your body, however. So, despite existing in comparable numbers to those cells, they account for only about 1.5 kilograms of the mass of a seventy-kilogram person.

The US government set up a five-year study, the Human Microbiome Project (*microbiome* refers to the microorganisms in a particular environment), or HMP, designed to identify all the foreign microorganisms in the human body and figure out what they do—a gargantuan task.[2] Its results were published in 2012 and revealed that there are more than 10,000 species of alien cells in your body—forty times the number of cell types that belong to you. In fact, every square centimeter of your skin is home to about five million bacteria. That is about five hundred in every pinhead-sized patch. The most densely populated regions of your skin are your ears, the back of the neck, the sides of the nose, and your belly button. What all these alien bacteria are doing is a mystery. Take bacteria in your nose, for instance. The HMP was at a loss to know what 77 percent of the species are doing there.

In fact, the HMP found that the nasal passages of 29 percent of people contain *Staphylococcus aureus*—better known as the MRSA superbug. This might sound worrying. However, in healthy people such bacteria are kept in check and cause no harm. It is only in sick people, whose immune systems are weakened, that MRSA bacteria are dangerous. Since hospitals contain sick people, it is obvious why MRSA can be a problem there.

There is mounting evidence that an imbalance of the human microbiome may be a factor in causing a wide range of diseases. These include inflammatory bowel conditions such as Crohn's disease and ulcerative colitis. There is even a suggestion that such imbalances may play a role in diseases such as Alzheimer's.[3]

Nobody is born with a full complement of foreign microorganisms. Instead, they are acquired after birth from a mother's milk and from the environment. All are pretty much in place by the time you are three years old. This is why it is possible to say that you are born 100 percent human but die 50 percent alien.

Actually, it is worse than this. The HMP found that the microorganisms that inhabit your body have a total of eight million genes, each of which codes for a protein with a specific purpose. By contrast, the human genome contains a mere 24,000 genes. Consequently, there are about four hundred times as many microbial genes exerting their effect on your body as human genes. Or, to put it another way, 97.75 percent of the DNA in your body does not belong to you. In a sense, you are not even as much as 50 percent human— you are a mere 0.25 percent human. Perhaps it is more truthful to say that you are born 100 percent human but die 97.75 percent alien!

6.

THE DISPENSABLE BRAIN

The juvenile sea squirt wanders through
the sea looking for a rock to cling to.
On finding one, it no longer needs
its brain. So, it...eats it.[1]

"The picture's pretty bleak, gentlemen...The world's
climates are changing, the mammals are taking over, and we
all have a brain about the size of a walnut."
—DINOSAURS IN GARY LARSON'S *THE FAR SIDE*

THE JUVENILE SEA SQUIRT wanders through the sea looking for a rock to cling to. On finding one, it no longer needs its brain. So, it...eats it. One wonders what a juvenile sea squirt does if it is dislodged from its chosen rock by a freak wave? Having eaten its brain, is it condemned to wander aimlessly through the ocean like a mariner without chart or compass? Or does it regenerate its thinking matter and miraculously recover the navigational skills required to locate a new rocky home?[2]

The autocannibalism of the juvenile sea squirt is a graphic illustration that brains are expensive in energy terms. This is why most creatures on Earth get by without one, and even a creature in possession of a brain—like the juvenile sea squirt—gets rid of it the moment it is no longer useful. Colombian-American neuroscientist Rodolfo Llinás puts it this way: "Basically there are two types of animals: animals, and animals that have no brains. They are called plants. They don't need a nervous system because they don't move actively, they don't pull up their roots and run in a forest fire! Anything that moves actively requires a nervous system; otherwise it would come to a quick death."[3]

Humans, of course, have an unusually large brain. And this is probably related to our switch to a diet that contains meat—which is more energy-rich than plant matter—and the invention of cooking. Just as writing acts as an external memory, the frying pan is an external stomach. It breaks down the proteins in meat so that they are easier to digest, doing some of the work the gut would normally do. With less energy needed to service a smaller stomach, there is more energy left over to support the needs of a bigger brain.

Incredibly, the human brain does all of its mega-computing on roughly twenty watts of power, which is the equivalent of a very dim light bulb. By comparison, a supercomputer capable of a similar rate of computation requires 200,000 watts—in other words, it is 10,000 times less energy efficient than the human brain. Despite the human brain's efficiency, however, it is extraordinarily energy hungry compared with all other tissues. While accounting for only 2 to 3 percent of the mass of an adult, it consumes about 20 percent of the body's oxygen.

The huge energy need of the human brain compared with the rest of the body is perhaps not surprising, given the fact that the brain contains something like one hundred billion brain cells—as many as there are stars in our Milky Way galaxy. Each brain cell, or *neuron*, may connect to 10,000 others via finger-like extensions called *dendrites*. This provides the potential for 1,000 trillion connections. And it is in these connections and their relative strengths that information such as memories are believed to be stored. Every experience you have, every moment, every day, changes the connectivity of your brain. In the words of American cognitive scientist Marvin Minsky: "The principal activities of brains are making changes in themselves."[4]

All this takes energy—which is why thinking hard makes you tired. "Whenever you read a book or have a conversation, the experience causes physical changes in your brain," says American science writer George Johnson. "It's a little frightening to think that every time you walk away from an encounter, your brain has been altered, sometimes permanently."[5]

But although the constant rewiring of the brain's neuronal connections takes a lot of energy, humans, unlike juvenile sea squirts, hang on doggedly to their brains.

Or so it seems.

The human brain, it turns out, has shrunk in mass by about 10 percent since it peaked in size 15,000–30,000 years ago.[6] One possible reason is that many thousands of years ago humans lived in a world of dangerous predators where they had to have their wits about them at all times to avoid being killed. Today, we have effectively domesticated ourselves and many of the tasks of survival—from avoiding immediate death to building shelters to obtaining food—have been outsourced to the wider society.[7] We are smaller than our ancestors too, and it is a characteristic of domestic animals that they are generally smaller than their wild cousins. None of this may mean we are dumber—brain size is not necessarily an indicator of human intelligence—but it may mean that our brains today are wired up differently, and perhaps more efficiently, than those of our ancestors.

We are at only the beginning of understanding how the human brain works. As far as science is concerned, it is the brain—not space—that is the final frontier. Some have suggested that it is a logical impossibility that the brain can ever fully comprehend the brain. "If the human brain were so simple that we could understand it, we would be so simple that we couldn't," pointed out American physicist Emerson M. Pugh.[8] However, the brain is of course not trying to understand the brain. Many brains are trying to understand the brain: the combined minds of the international scientific community. In the words of an Italian proverb: "All the brains are not in one head."

PART TWO

HUMAN
THINGS

7.

INTERACTION, INTERACTION, INTERACTION

There was no change in the design of stone hand axes for 1.4 million years

"Change is the only constant."
—HERACLITUS

THERE WAS NO CHANGE in the design of stone hand axes for 1.4 million years. I was told this extraordinary fact by Chris Stringer, an expert in human origins at London's Natural History Museum, a few years ago while researching my book *What a Wonderful World*. According to Stringer, palaeoanthropologists refer to the period as the "1.4 million years of boredom."

Of course, our hominin ancestors may have fashioned tools from wood, which quickly rotted in the ground, or from bone, making them nigh on impossible to distinguish from natural bone fragments.[1] And there is little doubt that profound changes were occurring in our ancestors' societies that left no obvious sign in the fossil record—such as the harnessing of fire, the invention of language, and the relentless rise in the complexity of social interactions.

However, the fact remains that, for about 60,000 generations, nobody thought of any improvement in the design of the stone hand axe. Today, by contrast, we live in a world where we expect the iPhone to be revamped next year, let alone in a generation's time. But rarely do we stop to reflect on how unprecedented the period in which we live is. For most of human history, nothing changed—or, if it did, it changed at a glacial pace (literally, since 90 percent of the past million years has been ice age).

What transformed everything was the invention of food production. Shortly after the end of the last ice age, about 13,000 years ago, humans began to experiment with cultivating crops. By 8,500 years ago, people in the Fertile Crescent of the Middle East had successfully domesticated wheat, peas, and olives, and by 8,000 years ago they had also domesticated sheep and goats. In China, pigs and silkworms were both domesticated by 7500 BC, along with rice and millet. With the advent of food surpluses, it was suddenly possible for people to live in moderate-sized settlements in which individuals had specialized "jobs" not connected with hunting or collecting berries.

By contrast, for most of human history, people lived in small bands of probably not much more than fifty. If anyone invented anything—maybe a design tweak on a stone hand axe—the invention is unlikely to have spread far beyond the group and would probably have died with them. Fire, for instance, may have been tamed many times and the secret repeatedly lost before becoming common knowledge.

But with the advent of large communities supported by the products of farming, it was possible for the first time for ideas and innovations to survive and spread. And with food production allowing the human population to grow everywhere, the opportunities for interaction grew remorselessly.[2] If there are three words that explain the history of the human race over the past 13,000 years they are: interaction, interaction, interaction.

The reason things are changing faster today than they did in the past is largely because there are more people—more opportunities for interaction, more chances for the virus-like spread of ideas and innovation—than ever before. And now, with the Internet, which enables social exchanges between billions upon billions of people, often geographically remote from each other, human interaction is skyrocketing.

This in itself is an amazing thing too. Just as the last 13,000 years have been unprecedented in human history, the last half a century or so has been unprecedented in terms of technological change. The reason for this is the doubling of computer power roughly every eighteen months, a trend first noticed in 1965 by Gordon Moore, one of the founders of the American computer-chip maker Intel and called "Moore's law" in his honour. Such a repeated doubling—which will make computers 1,000 times more powerful in fifteen years' time—cannot go on forever, of course. There are physical limits that will constrain how small and how fast the components of computers can be. We are therefore living through an extraordinarily unusual period of human history, the like of which we will never see again: the age of exponential growth of computing power. And we can barely guess what changes it will bring to human society.

But let's get back down to Earth—literally. The interaction, interaction, interaction that has created our modern world and makes it unlikely that the iPhone will remain unchanged for 1.4 years, let alone 1.4 million years, has been made possible by one thing and one thing only: farming. As an anonymous writer observed: "Despite his artistic pretensions, his sophistication, and his many accomplishments, man owes his existence to a six-inch layer of topsoil and the fact that it rains."

8.

THE GRANDMOTHER ADVANTAGE

Only three species have a menopause

"A study says owning a dog makes you ten years younger.
My first thought was to rescue two more, but I don't want to
go through the menopause again."

—JOAN RIVERS

MENOPAUSE HAPPENS WHEN A woman is depleted of her eggs, one or more of which is usually released from her ovaries every month. Because the total number of eggs released, or ovulated, in a woman's lifetime is about four hundred, typically the supply is exhausted by the time she reaches about fifty. (Incidentally, since your mother's eggs were present in her ovaries when she was an embryo in her own mother's womb, there is a sense in which you began your life not inside your mother but inside your grandmother.)

Remarkably, humans are one of only three species that are known to experience a shutdown of their reproductive potential before they die. The others are killer whales and short-finned pilot whales. (Spare a thought for female short-finned pilot whales—not only do they suffer the adverse symptoms of menopause but, if they get a hot flash, they have only short fins with which to fan themselves.)

What is so peculiar about a woman's reproductive potential shutting down well before death is that the winners in the Darwinian race are those whose genes are carried into the future by the largest number of offspring. Evolution would therefore appear to favor women who "get one more in" even as the final curtain comes into view. So why do women not release more than four hundred eggs? Why do they not release enough to last a lifetime?

Here, other factors may come into play. Later in life, not only is childbearing more risky but the chance of a child inheriting genetic defects is also higher. In addition, successfully rearing a child to adulthood takes a lot of energy. An older woman may lack the necessary energy and might even die, leaving a child motherless. Many biologists believe that, by switching off her ability to reproduce, a woman is then free to help her children rear children. This enhances the prospects of her grandchildren surviving, boosting the chance of her own genes getting into the next generation. It is all down to costs and benefits. The cost of her own pregnancy and the subsequent child-rearing set against the benefit of helping to rear a grandchild. Perhaps the latter wins out. Grandmothers do the unselfish thing, say biologists—ultimately out of selfishness.

However, not everyone agrees with the grandmother hypothesis in this precise form. Skeptics claim that the benefits for a grandmother who helps to rear her children's children, who share only quarter of her genes, do not outweigh the cost of giving up having her own children, who would share half of her genes.

Certainly in nonhuman species it is the case that females reproduce to a great age while reproduction is suppressed in their daughters, who help in the rearing of the infants. This makes sense since the daughters are ensuring the survival of individuals who share their genes. However, in most human societies, it is common for females to move away from their birth family. It makes no sense in evolutionary terms for them to help out their mother-in-law because any children will not share their genes. However, if a grandmother helps her daughter-in-law rear her children, this will help the grandmother get her genes into the next generation. "The mother-in-law's best strategy is to stop breeding, avoid competition, and allow the daughter-in-law to breed and help her," says Michael Cant, an evolutionary biologist at the University of Exeter in England.[1]

The grandmother argument is similar to the one used by scientists to explain the prevalence of homosexuality. Since the only way for genes and the characteristics they encode to propagate down the generations is through reproduction between a male and a female, genes that contribute to sex between same-sex couples should, by rights, become rapidly extinct. Yet the frequency of homosexuality is thought to be constant across cultures at about 3 percent in men and 2 percent in women.

One explanation for the persistence of homosexuality is that it has a genetic component that, though not in itself beneficial in promoting the cause of selfish genes, comes along with a gene that is. This is not uncommon in the natural world. There is a gene, for instance, that gives people immunity to malaria. However, if someone has not one copy of the gene—which is beneficial—but two copies, one from each parent, they develop sickle cell anemia, a cripplingly painful disease in which blood cells become flattened and block capillaries. Sickle cell anemia persists because, in most of the population, the gene that causes it has a beneficial effect and boosts people's chance of survival.

Of course, homosexuality may persist because homosexuals do get their genes into the next generation. Although there is a tendency to pigeonhole sexuality into a handful of distinct behaviors, it actually spans a whole spectrum, from 100 percent heterosexuality through bisexuality to 100 percent homosexuality. People may not be totally homosexual—or may be homosexual only at certain times in their lives. They might therefore have children. If this happens often enough, it would ensure that their genes persist and homosexuality does not die out.

But there is a more plausible way that homosexuals could get their genes into the future—and here there is the parallel with the grandmother explanation for menopause. If homosexuals help in the rearing of children who are genetically related to them—perhaps the offspring of a brother or sister—they will actually be acting selfishly to ensure their genes propagate into the future. This is also similar to the argument often employed by biologists to make sense of another great mystery of biology: altruism. Why do individuals do things that ensure the survival of others at the expense of their own survival? Again the theory goes that people are more likely to do that for people who are genetically related to them—that is, close family members. Selfishness explaining selflessness once more!

9.

LOST TRIBE

The crucial survival advantage that humans
had over Neanderthals was...sewing

"Despite being depicted in innumerable cartoons
as apelike brutes living in caves, Neanderthals had brains
slightly larger than our own. They were also the first
humans to leave behind strong evidence of burying
their dead and caring for their sick."

—JARED DIAMOND[1]

THE CRUCIAL SURVIVAL ADVANTAGE that humans had over Neanderthals may have been...sewing. This suggestion was made to me by Chris Stringer of the Natural History Museum in London. His point is that, although many bone needles fashioned by early humans have been found, no one has yet dug up a needle made by Neanderthals. Could it be that the ability to sew baby clothes during the bitterly cold ice-age winters gave human babies a few percent better chance of surviving than their Neanderthal peers, which is why Neanderthals died out around 40,000 years ago and we did not?[2]

The first Neanderthal fossils were discovered in Engis, Belgium, in 1829. But Neanderthals were not recognized as an early species of hominin until 1856 when a 40,000-year-old specimen was unearthed in a quarry in the Neander Valley near Düsseldorf in Germany.

Like other hominins, Neanderthals came originally from Africa. However, they migrated to Europe and Asia long before modern humans did. They lived in Europe and western Asia, as far east as southern Siberia and as far south as the Middle East. They survived from about 250,000 to about 40,000 years ago. For some of that time they lived alongside modern humans, who arrived on the scene about 130,000 years ago or possibly even earlier.[3] In fact, DNA evidence indicates that humans and Neanderthals shared a common ancestor about 600,000 years ago.[4]

Neanderthals have often been portrayed as numbskulls because their heads were big with a prominent brow ridge and large nose. However, their brains were actually slightly larger than ours; and the evidence suggests that they were far from stupid and to a large degree matched humans in their abilities. They made tools, cooked their food, created artworks, and buried their dead in a ritualized manner.[5] They were people, just like us. Neanderthals, however, were stockier and more muscular than humans. This is believed to be an adaptation for survival in the harsh ice-age environment, since the more compact a person's build the smaller the surface area of their skin through which precious heat can escape.

The last Neanderthals disappeared about 40,000 years ago.[6] Why they did so is one of the outstanding questions of palaeoarchaeology. The theory that Neanderthals were unable to sew baby clothes is one among many other possibilities. Neanderthals may have not been as good at adapting to changes in climate as modern humans, or they may simply have been outcompeted by modern humans for territory and game.

The mystery is compounded by the fact that Neanderthals were successful for a very long period of time in ice-age conditions, which were extremely harsh. It seems they were adept at ambushing and killing very large prey, such as mammoths, bison, and reindeer. But for some reason this highly successful species—our closest ancient human relative—disappeared from the face so recently, it amounts to a mere blink of an eye in the history of hominin evolution.

However, there is a twist to the story. It turns out that up to 2 percent of the DNA of people living outside of Africa is Neanderthal.[7] So, Neanderthals did not become entirely extinct: they interbred with humans. At this moment, they are walking among us.

10.

MISSED OPPORTUNITY

There is no photograph of the first
man on the moon

"When I first looked back at the earth,
standing on the moon, I cried."
—ALAN SHEPARD (*APOLLO 14* ASTRONAUT)

URING THE APOLLO PROGRAM, NASA, by its own estimates, spent about $25 billion on sending humans to the moon, which must be well in excess of $100 billion dollars in today's money. But, incredibly, the American space agency missed out on arguably the biggest photo opportunity of all time. It failed to obtain a photograph of the first man on the moon, Neil Armstrong. Buzz Aldrin, Armstrong's copilot and the second man to set foot on the moon, simply did not take one.

Actually, this is not quite true. There is one photo of Armstrong on the moon—only he has his back to the camera![1] Additionally, in Armstrong's iconic photo of Aldrin on the lunar surface, it is possible to see his own small white figure reflected in Aldrin's helmet visor. And, of course, there are the fuzzy black-and-white TV images that were beamed back to Earth of Armstrong and Aldrin's stopover on the moon. But that is it. The first human to stand on another world—arguably as significant an individual as the first fish to crawl from the sea on to the land—was essentially unphotographed.

Aldrin was not actually to blame for this. For most of the two hours and thirty-one minutes he and Armstrong were out and about on the lunar surface, Armstrong was scheduled to have the camera. It was a Hasselblad Electric Data Camera, specially adapted from the motorized Hasselblad 500 EL.[2] It used seventy-millimeter film and a polarizing filter. The astronauts carried the camera mounted on their chest. It had no viewfinder, and they had to guess what was visible through the lens.

Before the Apollo 11 moon mission, Armstrong and Aldrin had each been given a camera to practice with at home. Nevertheless, the surface of the moon poses unique challenges for photography. Whereas on Earth sunlight is scattered by air molecules, softening its harshness and spreading it even into the shadows so they are not completely black, there is of course no air on the moon. A camera must contend with dazzling brightness alternating with utter blackness, the two zones separated by a knife-sharp boundary. Not an easy thing for an exposure meter to contend with.

But it is not simply the stark contrast between light and shadow that makes the lunar surface appear so weird. On Earth, the scattering of light by dust in the air causes distant objects to appear faded and blurred. This gives us vital and mostly unconscious clues about what things are near and what things far away. On the moon, which lacks an atmosphere, there are no such clues. It is impossible to distinguish between a 20-meter hillock 20 meters away and a 2-kilometer mountain 2 kilometers away. There is an alienness that is difficult to appreciate in mere photographs.

Another property of the lunar surface that the images have failed to convey is its colors. Far from being grey and dull, the landscape is painted in shimmering silvers and browns and golds—colors that, peculiarly, change depending on a camera's point of view. This is all down to moon dust. Whereas grains of sand on a beach are incessantly tumbled and rubbed together, polishing and smoothing them into miniature smooth pebbles, no such process operates on the moon. Instead, the lunar surface is bombarded by a constant rain of micrometeorites. These tiny specks of dust impact at such high speed that they shatter and heat the rock, creating dust grains more reminiscent of melting snowflakes than pebbles. It is the jagged nature of moon dust, which causes it to reflect sunlight very differently in different directions, that makes its color appear different from different viewpoints.

Once upon a time, there was a real fear that parts of the moon were covered in such a deep layer of dust that a spacecraft would sink without trace. In Arthur C. Clarke's 1961 novel *A Fall of Moondust*, for instance, the lunar dust cruiser *Selene* sinks with all its passengers in a sea of lunar dust. Fortunately, fears of lunar quicksand proved unfounded.

According to the Apollo crews, moon dust smelled of gunpowder. It snagged on their spacesuits and made them look more like coal miners than astronauts. Sadly, the only geologist to go to the moon— Harrison Schmitt, on the final mission, Apollo 17—was allergic to moon dust.[3] He must have sneezed all the way home!

The constant battering of the lunar surface by micrometeorites causes the lunar "soil" to turn over in about ten million years. This process of "lunar gardening" means that, although the footprints left by the astronauts will probably outlast the human race, they will not last forever.

Armstrong and Aldrin left their footprints in the dust of the Sea of Tranquility on July 20, 1969. About 3.6 million years before that, a small band of hominins left their footprints in the volcanic dust of Laetoli, Tanzania.[4] Surely there cannot be a more striking illustration of how far humans have come—and how much we stand to lose if we find no solutions to the global problems that threaten human extinction.

PART THREE

TERRESTIAL THINGS

11.

THE ALPHABET OF NATURE

Every breath you take contains an atom
breathed out by Marilyn Monroe

"If in some cataclysm, all of scientific knowledge were to
be destroyed, and only one sentence passed on to the next
generations of creatures, what statement would contain the most
information in the fewest words? Everything is made of atoms."
—RICHARD FEYNMAN[1]

IN ABOUT 440 BC, the Greek philosopher Democritus picked up a
rock or a branch (or it may have been a shard of pottery) and asked
himself the question: if I cut this in half, then in half again, can I go
on in this way forever? He was in no doubt about the answer. It was
utterly inconceivable to him that matter could be subdivided forever.
Sooner or later, he reasoned, a grain of matter would be arrived at
that could not be cut in half. Since the Greek for "uncuttable" was
a-tomos, Democritus called such indestructible grains *atoms*.

Democritus's atoms were actually more than indivisible grains of
matter. He also postulated that they came in a small number of different types. By combining them in different ways, it was possible to
make a rose or a chair or a newborn baby. "By convention there is
sweet, by convention there is bitterness, by convention hot and cold,
by convention color," he wrote. "But in reality there are only atoms
and the void."

This is an extraordinarily powerful idea. It recognizes that the complexity of everything around us is in fact an illusion. Beneath the skin of the world, everything is simple. Complexity is created merely by combining a handful of basic building blocks in an infinity of different ways. Everything is in the combinations. Atoms are the Lego bricks of nature.

Democritus's idea that nature, at its fundamental level, is simple turns out to be the act of faith that drives modern science. Nobody knows why the universe should be simple underneath. But the proof of the pudding is in the eating. And, in the past few centuries, scientists have had continual success in finding evermore simple, basic laws of physics that underpin the world.

Democritus knew that atoms must be very small because they were invisible to the naked eye. We now know, in fact, that it would take ten million of them laid end-to-end to span the full stop to the end of this sentence. By the beginning of the twentieth century, there was lots of indirect evidence of the existence of atoms. For instance, the pressure exerted by a gas can be explained if myriad tiny grains are drumming on the walls of a container like raindrops on a tin roof. However, it was not until relatively recently that anyone actually saw an atom.[2]

In 1980, Heinrich Rohrer and Gerd Binnig, working at IBM in Zürich, built a scanning tunneling microscope, or STM. In exactly the same way that a blind person can build up a picture of someone's face by dragging their finger across their skin, an STM builds up a picture of the microscopic atomic terrain of a surface by dragging a tiny stylus across it and recording the up-and-down motion. In the computer images created from this information, atoms look like tiny footballs, like oranges stacked in boxes...in fact, how Democritus imagined them more than two millennia ago.

For their invention of the STM, Rohrer and Binnig won the 1986 Nobel Prize for Physics. Of course, there were no Nobel Prizes in Democritus's day. But he surely deserves to be in *Guinness World Records* as the person who made a prediction the most in advance of its confirmation.

But here's the point of all this: every time you take a breath, you suck in a certain volume of air. For simplicity, let's call that volume a mouthful. How many mouthfuls stacked together would be equal to the volume of the earth's atmosphere? Obviously, a very large number. However, it turns out there are more atoms in a single mouthful of air than there are mouthfuls of air in the earth's atmosphere. It thus follows that every breath you take contains atoms breathed out by Marilyn Monroe. Or Julius Caesar. Or the last Tyrannosaurus Rex ever to have stalked the earth.

12.

ROCK SPONGE

When the tide at sea rises,
the water in wells falls

"There is a tide in the affairs of men, which taken at the
flood, leads on to fortune. Omitted, all the voyage of their
life is bound in shallows and in miseries."
—WILLIAM SHAKESPEARE, *JULIUS CAESAR*

D ID YOU KNOW THAT the water in a well drops when there is a
high tide at sea and rises at low tide? The phenomenon has been
known since about 100 BC when Greek philosopher Poseidonios
noticed it on the Atlantic coast of Spain. His original observations
are lost. However, they are reported by the Greek geographer Strabo
in his *Geographica*: "There is a spring at the [temple of] Heracleium
at Gades [Cadiz], with a descent of only a few steps to the water
(which is good to drink), and the spring behaves inversely to the flux
and reflux of the sea, since it fails at the time of the flood-tides and
fills up at the time of the ebb-tides."

Incredibly, Poseidonios's peculiar observation was not explained
for two millennia. Only in 1940 did the American geophysicist
Chaim Leib Pekeris realize it was a consequence of the fact that the
moon creates tides not only in the earth's oceans but in its rocks as
well (to be precise, the tides are caused by both the moon and the
sun, with those of the moon being twice as big as those of the sun).

As Isaac Newton first understood, tides are a distortion in the shape of the earth caused by the pull of the moon being stronger on parts of the earth closer to it. Imagine the ocean immediately beneath the moon. The pull on water at the surface is greater than the pull on water at the seabed. The difference causes the sea to bulge upwards towards the moon. A similar effect causes a second tidal bulge on the opposite side of the earth to the moon. And, as the earth rotates through the two bulges, any location gets two tides a day, as the ocean is pulled upwards and slumps back down again.

But the moon also causes a tidal bulge in the solid earth. It is less noticeable since rock is a lot stiffer than water. The explanation of Poseidonios's observation is the following: the ground around a well is, not surprisingly, waterlogged. Think of it as a wet sponge. When the sponge is stretched upwards at high tide, it sucks water out of the well, lowering its level; and, when the sponge is scrunched back down at high tide, it squeezes water back into the well, raising its level.

A more contemporary example of such rock tides comes from the 26.7-kilometer subatomic racetrack of the Large Hadron Collider (LHC) near Geneva. Around this ring, counter-rotating beams of protons are smashed together at 99.999999 percent of the speed of light. In July 2012, they created the fabled Higgs particle, the "quantum" of the Higgs field, which endows all other subatomic particles with mass.

The LHC occupies the same tunnel as an earlier accelerator known as the Large Electron-Positron Collider (LEP). And in 1992 physicists at LEP noticed a peculiar thing. Twice a day, they had to adjust the energy of the circulating electrons and positrons to keep them in the tunnel. It seemed that twice a day the circumference of the ring was changing by about 1 millimeter. After scratching their heads for a while, the physicists realized why: the rock beneath the ring was being stretched and squeezed by the moon!

13.

DEEP IMPACT

When an asteroid wiped out the dinosaurs,
they got less than ten seconds' warning

"The dinosaurs became extinct because they
didn't have a space program."
—LARRY NIVEN

THE CITY-SIZED ASTEROID THAT slammed into the earth and wiped out the dinosaurs sixty-six million years ago was not only small but, most likely, as black as coal.[1] There would have been no sign in the sky of its approach until it hit the top of the atmosphere and friction with the air caused it to heat up to incandescence. Traveling at around seventeen kilometers per second, it would have taken less than ten seconds to reach the ground. That's how much warning the dinosaurs got.

Extraterrestrial impacts on Earth are surprisingly common. In 1908, a body the size of a couple of houses exploded about five kilometers above the Tunguska River in Siberia, flattening 2,000 square kilometers of forest. It was equivalent to about 1,000 Hiroshimas. More recently, in 2013, a similar body disintegrated high above Chelyabinsk in Russia with the explosive power a seven-megaton H-bomb.[2] But, although such bodies are expected to hit the earth about once a century and bodies about a kilometer across once every half a million years, asteroids the size of the one that hit sixty-six million years ago are, thankfully, expected to come our way only once every one hundred million years.

The first evidence of a catastrophic impact around this time came from a team led by American physicist Luis Alvarez. In 1980, it reported the discovery of a layer of the element iridium, which occurs all across the globe in strata about sixty-six million years old.[3] Iridium is not abundant on the surface of the earth but is more common in extraterrestrial material such as meteorites. Alvarez's team therefore concluded that an asteroid impact sixty-six million years ago had finished off the dinosaurs.

The case was later bolstered by the discovery of large, 180-kilometer diameter, semi-submerged crater at Chixculub off the coast of Yucatán in Central America. The recovery of shocked grains from the site confirmed that there had been an impact, and the date appeared to match the date of the iridium layer.

Yet puzzles remain. Before the impact, the dinosaurs had been in decline for millions of years. This was probably connected with the mega-eruption of the Deccan Traps in present-day India, which pumped out lava, in some places two kilometers deep, over an area of about 500,000 square kilometers. Sulphur dioxide associated with the eruption and injected into the atmosphere would have reflected sunlight back into space, cooling the planet and stressing the ecosystem of the dinosaurs. It seems, therefore, that dinosaurs were on their way out sixty-six million years ago and the Chixculub impact was merely the coup de grâce.

Peculiarly, although most species of dinosaur were wiped out—the exception being the line that led to modern-day birds—amphibians, which are often considered barometers of environmental degradation, survived.[4] Precisely what killed the dinosaurs after the impact of the asteroid is still a matter of debate—the impact released a million times the energy of the biggest H-bomb, but it was localized. One theory is that, as it struck the ocean, it caused a mega-tsunami. And metals like nickel, which are common in asteroids, would have created a toxic rain. However, interestingly, recent research suggests that the dinosaurs were simply unlucky.

The impact site at Chixculub was rich in hydrocarbons such as oil, which is true of only about 13 percent of the surface of the earth. These would have been ignited by the impact, injecting a plume of sooty smoke into the stratosphere.[5] High above the weather, this material would not have been rained out for years. It would have cloaked the earth, screening out sunlight and plunging the planet into a deadly winter.

Now, for the first time in history, we recognize the danger of extraterrestrial impacts. Surveys have logged tens of thousands of bodies whose orbits cross that of the earth and which could potentially cause disaster. However, we lack the technology to do much about an asteroid that has our name written on it. A space mission to shatter an asteroid would merely create myriad smaller pieces, all heading towards us on the same deadly trajectory. A better strategy would be to land on an asteroid. If material were ejected in one direction, the asteroid could be pushed in the opposite direction. Over many months or years, this rocket effect could nudge the body's trajectory, shifting it from harm's way. Since such a space technology is still over the horizon, we will simply have to keep our fingers crossed that we are not as unlucky as the dinosaurs. And ponder what we would do if we suddenly realized we had only ten seconds to live!

14.

SECRET OF SUNLIGHT

Contrary to expectations, the earth
does not have an energy crisis

"Only entropy comes easy."
—ANTON CHEKHOV

HOW MUCH ENERGY DOES the earth trap from the sun? The answer is, surprisingly, none. All of the solar energy intercepted by our planet is radiated back into space.[1] Were this not the case, the earth would simply get ever hotter until its surface became a molten goo.

So, if it is not solar energy that is ultimately powering every living thing on Earth, not to mention our global technological civilization, what is it? The answer to this is useable solar energy—and the distinction between the two is both subtle and profound. In fact, our understanding of how the earth exploits the sun's energy and ultimately spits it back involves one of the most important themes in physics: thermodynamics.

The first thing to consider is the fact that the particles of light, or photons, from the sun are typical of the sun's surface temperature of about 5,500 °C, whereas the photons that are re-radiated into space by the earth are typical of the earth's much lower average surface temperature of about 20 °C.

A direct and meaningful comparison of the energy of these photons requires using the Kelvin temperature scale. On the scale, 0 Kelvin—the lowest possible temperature—corresponds to -273 °C.[2] Whereas the sun's photons correspond to about 5,800 Kelvin, the earth's correspond to around 300 Kelvin. This means that the photons radiated by the earth have 300/5,800—or roughly one twentieth—the energy of the ones coming from the sun. Given that no net energy is absorbed by the earth, this implies that the earth radiates twenty low-energy photons for every high-energy one it receives from the sun.

It is easier to keep track of one thing than many things. Intuitively, therefore, a single photon is a simpler, more ordered thing than twenty photons. And ordered sources of energy, it turns out, are better than disordered sources at making things happen—or, in the physics jargon, of doing "work." In other words, each photon received from the sun is better at doing work than the twenty photons the earth radiates in return.

At this point it is necessary to make a small excursion into the realm of steam engines, which, contrary to expectations, are not simply the machines that powered the nineteenth-century Industrial Revolution but are devices of truly universal significance. "All of our actions, from digestion to artistic creation, are at heart captured by the essence of the operation of a steam engine," says chemist Peter Atkins.[3] The reason is that steam engines illustrate at a fundamental level how energy does its work—they do this so well, in fact, that the subject of thermodynamics initially developed through our study of them.

In a steam engine, steam at high temperature pushes a movable wall, or "piston," against the pressure of the air. Having done its work, the steam condenses into water at low temperature. Basically, something at high energy (in this case, high temperature) does work and ends up at low energy (low temperature).

Temperature is actually a measure of random microscopic motion. The molecules of steam (water) are flying around very fast in all directions like a swarm of very angry bees, which is why the steam has a high temperature. In battering the piston and pushing it forward, molecules transfer some of their random motion to the bulk motion of the piston and therefore slow down. This is why the temperature drops and the steam ends up as water.

You might think all the heat energy of the steam can be turned into work. But you would be wrong. This is because only some of the molecules will be moving in the direction of the piston, enabling them to push it forward, whereas many will not be. This is a fundamental result: energy can never be converted into useful work with 100 percent efficiency.

The amount of energy that can do useful work is called the exergy. And it turns out that energy at low temperature has less exergy than the same amount of energy at high temperature. It is degraded, sapped of the ability to do anything. This applies whether it's the condensed water in a steam engine or the photons radiated by the earth into space. In a nutshell, what happens on Earth is that the sun's high-exergy photons are received and set to work in countless biological and technological processes—each, at a fundamental level, equivalent to a small steam engine—which drain their energy of its ability to do such useful work, then finally radiate them back into space in a far more useless, low-exergy form.

You may have heard of the term *entropy*. This is a measure of the amount of disorder of a system such as a volume of steam. Think of it this way. Heat energy at high temperature is like a noisy restaurant. Adding energy is like standing in the doorway and shouting to a diner. The shout goes pretty much unnoticed—that is, it does not increase the noisy disorder much. Heat energy at low temperature, however, is like a quiet library. Adding the same amount of energy as you did to the restaurant is like standing in the doorway and shouting to a reader. The same shout upsets the calm of the library. It greatly increases the disorder.

Well, every time heat energy does work in a steam engine (or in one of the earth's innumerable processes), the result is heat at a lower temperature, which is more disordered, or has higher entropy. And disordered heat has less ability to do anything useful. It is the flip side of exergy (energy with high entropy has low exergy, and vice versa).

So, to return to that important difference between solar energy and useable solar energy: the earth may not use any net energy from the sun, but energy from the sun does do work on Earth—just like energy does work in a steam engine. And every time it does, that energy's capacity to work is depleted. In essence, solar energy is worked to death before finally being discarded, exhausted back into space.

PART FOUR

SOLAR SYSTEM & THINGS

15.
CELEBRATING MASS

If the sun were made of bananas it would not
make any difference

"The sun is a mass of fiery stone,
a little bit bigger than Greece."
—ANAXAGORAS, 434 BC

I F THE SUN WERE made of bananas, it would not make any difference. Well, not much difference. It is all down to why the sun is hot.
The sun is hot for an incredibly simple reason: it contains a lot of
mass. All that matter bearing down on the solar core squeezes the
material there. And, when things are squeezed, they get hot, as anyone who has squeezed the air in a bicycle pump knows. The temperature at the center of the sun is about fifteen million degrees.
And, at such a temperature, matter dissolves into an amorphous,
anonymous state known as a plasma. It does not matter what the
matter is, it always ends up in this state.

Now, the sun is about a billion billion billion tonnes of mostly
hydrogen gas. But, if you put a billion billion billion tonnes of microwave ovens in one place, or billion billion billion tonnes of bananas, you would get something equally as hot as the sun. The point
is that the temperature of the sun is determined essentially by the
amount of matter it contains and not by the type of that matter.

But this explains only why the sun is hot at this instant not why it stays hot. The sun is continually losing heat into space, yet its temperature does not change noticeably. This tells us that something must be replacing the heat exactly as fast as it is lost.

In the nineteenth century, in an era powered by steam, it was natural to think that the sun was a lump of coal. Of course, it would have to be the mother of all lumps of coal. The problem was that, according to the calculations of the Scots-Irish scientist Lord Kelvin, a coal-powered sun would burn out in only about 5,000 years. This was not even enough for Irish Archbishop James Ussher, who, from a detailed analysis of the Bible, calculated that the earth (and consequently the sun) was born on October 23, 4004 BC at 9:00 a.m. And it was certainly not enough for the geologists and biologists.

Geologists had discovered fossil sea creatures on mountaintops and deduced, quite reasonably, that such mountains must have begun their lives beneath the ocean. Since nobody had ever seen a mountain grow in a human lifetime, it was clear that tens of millions of years would be required for one to reach its full height. And even longer expanses of time were needed by the biologists. Charles Darwin had provided abundant observational evidence that today's profusion of living things had diverged from a simple common ancestor by a process of natural selection. Since nobody had ever seen one species morph into another in a human lifetime, it was clear that Darwin's painstakingly slow process required hundreds of millions, if not billions, of years.

In fact, *radioactive dating* of meteorites—the builders' rubble left over from the birth of the solar system—showed that the earth and, by implication, the sun is 4550 million years old. In other words, the sun has existed for about a million times longer than would be possible were it a lump of coal. Or, to put it another way, whatever is powering the sun is an energy source about a million times more concentrated than coal. At the beginning of the twentieth century, remarkably, such a source came to light: nuclear energy.

The sun is fusing the cores, or nuclei, of hydrogen, the lightest of atoms, into the second lightest, helium, the mass difference between the initial and final products appearing as the energy of sunlight, according to Einstein's famous formula $E = mc^2$. Because of this process, the sun gets lighter by the mass equivalent of one million elephants every second (by comparison, the biggest H-bomb turned only about a kilogram of mass-energy into other forms of energy, chiefly heat).

The sunlight-generating nuclear reactions in the sun are extremely sensitive to temperature, slowing down if the sun cools and speeding up if it heats up. But, if they generate too much heat, the gas of the sun, like any gas that is heated, expands and cools, throttling back the nuclear reactions; if they generate too little, the gas shrinks and heats up, revving up the nuclear reactions. The sun therefore has a built-in thermostat. The upshot is that the nuclear reactions maintain the sun at precisely the temperature determined by its mass alone. Incredibly, then, the temperature of the sun has nothing to do with the details of its energy source (which is why it has the same central temperature it would have if it was made of bananas).

The first step in assembling a helium nucleus is for two hydrogen nuclei in the heart of the sun to slam into each other and stick together. On average, this process takes about ten billion years, which is why the sun will shine for about ten billion years, meaning it is barely halfway through its life. This nuclear reaction the sun uses is just about the most inefficient nuclear reaction imaginable. Picture your stomach and a volume of the core of the sun the size and shape of your stomach. Your stomach generates energy at a faster rate! You might then wonder how the sun stays so hot. The answer is that the sun does not consist of one chunk of matter the size and shape of your stomach but rather countless quadrillions of such chunks all stacked together.

Next time your skin is being warmed by the sun on a summer's day, be thankful for the mind-boggling inefficiency of solar-nuclear reactions. Their slowness is the reason the sun has been able to shine for the billions of years necessary for the evolution of complex life like you.

16.

KILLER SUN

Once upon a time, people on Earth were
electrocuted by a solar flare

"If there is a solar flare or a nuclear war, a thousand cans of
pickled turnips aren't going to save you."
—SARAH LOTZ[1]

ACROSS THE WORLD, TELEGRAPH operators were electrocuted and, at low latitudes, a blood-red aurora borealis appeared, so bright a newspaper could be read by it at midnight.[2] The Carrington event, named after amateur astronomer Richard Carrington, who from south of London noticed a flare on the sun at the same time that a magnetometer at Kew flew off-scale, changed forever our ideas about the sun.[3] Before September 1, 1859, our local star was believed to influence the earth only through its gravity and of course the warming effect of sunlight. Afterwards, it was realized that violent convulsions on the solar surface, or photosphere, could fire magnetic missiles at our planet with devastating effects.

In the 1920s, the British astrophysicist Sir Arthur Eddington deduced the internal structure of the sun and its central temperature of more than ten million degrees merely by assuming that it is a giant ball of gas. The key to this was his recognition that, because the sun is not noticeably expanding or contracting, every portion of its interior must be in perfect balance. In such a state of "hydrostatic equilibrium," the force of gravity pulling inward on every chunk of solar matter must be perfectly countered by the force of the hot gas pushing outwards. Although we now know that the sun's heat comes from the nuclear fusion of hydrogen to helium, the by-product of which is sunlight, remarkably Eddington's conclusions did not require him to know anything about the source of solar heat. As pointed out in Chapter 15, the sun's central temperature depends essentially only on its mass and would be the same for a similar mass of bananas, rusty bicycles, or discarded TV sets.

Eddington's sun was a predictable and somewhat dull ball of hot gas. However, the fact that it has a magnetic field changes everything. It makes the nearest star an unpredictable, seething, explosive, infinitely surprising laboratory for extreme physics.

Magnetic fields are generated by moving electric charges. In the case of a mundane bar magnet, the movement is only of the electrons inside atoms, and the atoms stay put. The key thing to understand about the sun is that it is not an ordinary gas—it is an electrically charged gas, or plasma, of nuclei and electrons. And, in the solar plasma, the moving charges that create the magnetic fields are free to move, unlike the atoms in a bar magnet. This movement changes the magnetic field, which in turn affects the movement of the charges, which changes the magnetic field again, and so on...It is this complex interplay between the hot plasma and magnetic field that is behind all the myriad solar-magnetic phenomena from the magnetic whorls of sunspots to the mega-detonations of solar flares.

Actually, there is one other essential ingredient. The sun is not a rigid body. Its exterior rotates at a different rate to its interior, and even regions of its exterior at different latitudes rotate at different speeds. Consequently, the magnetic fields in the sun are continually being twisted and contorted, storing up energy just like twisted-up elastic bands.

Where loops of magnetic field break through the surface, we see sunspots—nearly always in pairs since the loop emerges from the sun at one point and reenters at another. Where a magnetic field becomes so twisted it snaps, reconnecting with other fields, the energy unleashed hurls million-degree plasma tens of thousands of kilometers above the sun in a solar flare. There is even a million-mile-an-hour hurricane—the solar wind—that blows from the sun, carrying its magnetic field out through the solar system. In a very real sense, the earth orbits inside the atmosphere of the sun. In fact, that atmosphere comes to an end only way beyond the outermost planet, where the solar wind slams into the interstellar medium like a snow plough running into a snow drift. On August 25, 2012, NASA's *Voyager 1* space probe, launched in 1977, detected a strong increase in *cosmic rays*, high-energy particles from our galaxy, making it the first human vehicle to leave the sun's atmosphere and actually taste interstellar space.

Understanding the sun is more than a mere academic activity. Our very survival on Earth may depend on predicting the space weather created by the nearest star. Studies of other sun-like stars reveal that they can launch mega-flares—though admittedly rarely—that are quite capable of frying a planet like the earth. A more serious concern is Coronal Mass Ejections (CMEs), which should more accurately be called "coronal magnetic eruptions." First recognized in the 1970s, these are missile-like ejections into space of vast amounts of solar plasma and magnetic fields. We are talking about something of roughly the mass of Mount Everest hurled into space at five hundred times the speed of a passenger jet. The Carrington event—the most violent solar event ever recorded—is now recognized to have been a CME.

In 1859 the world was not dependent on electrical technology, so the CME caused no serious harm to civilization. Contrast this with the situation today. Changes in the magnetic field across electrical power grids can induce currents big enough to melt equipment. Such *induction* was behind the electrocution of telegraph operators in 1859 and a major power outage in Quebec on March 13, 1989 which left six million people in the dark.[4] But the real threat today is to the dense ring of satellites that girdles our planet and on which our lives now depend. Communications satellites, weather satellites, Global Positioning System (GPS) satellites—which not only allow us to know our location but play a crucial role in global financial transactions—are all at risk. In rich countries, efforts have been made to harden infrastructure against future CMEs. However, it is a sobering thought that the sun, which has given us life, could in an instant return us to a preelectrical age.

17.

LIGHT OF OTHER DAYS

Today's sunlight is 30,000 years old

"Every time four protons are turned into a helium nucleus, two
neutrinos are produced. These neutrinos take only two seconds
to reach the surface of the sun and another eight minutes or so
to reach the earth. Thus, neutrinos tell us what happened in the
center of the sun eight minutes ago."
—RAYMOND DAVIS

"Fond memory brings the light of other days around me."
—THOMAS MOORE

ONE OF THE MOST amazing images in the history of science is
a grainy yellow-orange blob on a background of dark blue.
When I give public talks, I project it on a big screen and ask people
what they think they are looking at. Answers range from an explod-
ing star to an atom to a ball of molten metal. If nobody guesses the
correct answer—which, thankfully, is usually the case—I say, as dra-
matically as I can: "It's an image of the sun…taken at night."

"Wait a minute," someone will invariably say. "Surely, at night, the
sun is below the horizon?"

"Exactly. The image was taken not looking up at the sky, but looking down through 8,000 miles of solid rock to the sun on the far side
of the earth. Not with light but with neutrinos."

Neutrinos are ghostly subatomic particles produced in prodigious
numbers by the sunlight-generating nuclear reactions in the heart of

the sun. Hold up your thumb. About one hundred billion neutrinos are passing through your thumbnail every second. You never notice them because the key characteristic of neutrinos is that they are fantastically antisocial. They hardly ever interact with the atoms of everyday matter. The only way of registering them is to build a detector containing a large number of atoms, thus boosting the chance of at least one of those atoms stopping a neutrino.

The image of the sun seen through the earth was made by the Super-Kamiokande neutrino detector, which is buried in a cavern deep inside the Japanese Alps. Imagine a giant baked-bean can ten stories high filled with 50,000 tonnes of water. Very occasionally, a solar neutrino traveling through the detector interacts with a hydrogen nucleus, or proton, in a molecule of water. The resulting subatomic shrapnel explodes through the water, creating the light equivalent of a supersonic shockwave. More than likely you have seen a picture of such "Cherenkov light." It is the blue glow that comes from radioactive waste stored in ponds at the sites of nuclear reactors.

The inside of Super-Kamiokande's giant baked-bean can is covered with what look like light bulbs, fifty centimeters in diameter. There are 11,146 of these light detectors, or "photomultiplier tubes." And by noting which ones are triggered and in what order, the physicists can deduce the path of a neutrino that creates Cherenkov light.

But the details of the detector are not important. The point is that neutrinos interact so rarely with matter that they are unimpeded as they travel out of the sun. Their path from the heart of the sun to its surface is therefore a straight line. It takes just two seconds. Once at the surface, the neutrinos take about 8.5 minutes to fly across space to the earth.

Hold up your thumb again. About 8.5 minutes ago the neutrinos passing through it were in the center of the sun.

Now, contrast the neutrinos with light, which is also created by the nuclear reactions in the heart of the sun. Light, which is a stream of countless bullet-like particles, has tremendous difficulty getting out the sun. Photons are like Christmas shoppers fighting their way down a crowded street. They cannot go in a straight line but are forced to zigzag. In the sun, they never travel more than a centimeter before being deflected in another direction. In fact, their path out of the solar interior is so tortuous that it takes not two seconds, as it does for neutrinos, but about 30,000 years. Thereafter, it takes 8.5 minutes to fly to the earth.

Consequently, today's sunlight is about 30,000 years old. It was created at the height of the last ice age.[1]

18.

A BRIEF HISTORY OF FALLING

Though it does not look like it, the moon is
perpetually plummeting towards the earth

"The knack of flying is learning how to throw yourself
at the ground and miss."
—DOUGLAS ADAMS[1]

WHY DON'T SATELLITES FALL down? More than once I have
been asked this question by schoolchildren. The answer, surprisingly, is that they *are* falling down—but they never reach the
ground! The first person to realize this far-from-obvious fact was
Isaac Newton in the seventeenth century. He was of course not
thinking about artificial satellites but rather the earth's natural satellite: the moon. Newton wondered why the moon orbits the earth,
and he came up with the following explanation.

Imagine a cannon that fires a cannonball horizontally across the
ground. Gravity causes its trajectory to curve downwards and after
perhaps one hundred meters it hits the ground. But imagine a much
bigger cannon that fires a much faster cannonball. It travels further—maybe a kilometer—before its downward-curving trajectory
intercepts the earth. Now, imagine a gigantic cannon—the mother
of all cannons—which fires a cannonball at 28,080 kilometers an
hour. At this enormous speed, the fact that the earth is a ball be-

comes of key importance. As fast as the cannon ball curves down towards the earth, the earth's surface curves away from the cannon ball. The cannon ball therefore never hits the ground. It falls forever in a circle.

This, reasoned Newton, is what the moon is doing—perpetually falling in a circle.[2] Proof that he was right is today provided by the International Space Station. Surprisingly, gravity at the altitude of its orbit is about 90 percent of what it is on the surface of the earth. The astronauts on board are therefore weightless not because they are beyond gravity, as is commonly imagined, but because they are falling. And, a person in free fall feels no gravity.[3] If you are ever unlucky enough to be in a lift whose cable snaps, you will realize the truth of this statement.

The moment Newton realized that the moon is falling, he was able to make a courageous leap of the imagination. Whereas the prevailing opinion of his time was that there exists one set of laws that govern the earth and another set that govern the heavens, Newton claimed that both heaven and Earth are governed by exactly the same laws. In particular, he maintained that the same law of gravity that causes an apple to fall to the ground from a tree causes the moon to fall towards the earth.

By comparing the two forces, he was able to determine the behavior of his universal law of gravity. He discovered that it obeys an *inverse-square law*. In other words, the force of attraction between two masses becomes four times as weak if their separation is doubled; nine times as weak if it is tripled; and so on.

Newton deduced his universal law of gravity in his "miraculous year," which spanned 1665 and 1666. With plague threatening Cambridge, he fled the university for the safety of his family home in Woolsthorpe, Lincolnshire. There, in enforced exile, he not only penetrated the mystery of gravity but also discovered that sunlight is made of all the colors of the rainbow and invented the mathematics of calculus.

Bizarrely, though, Newton did not tell anyone about his law of gravity for almost twenty years. What finally prompted him to go public was an approach made to him by Edmund Halley, today best known for the comet that bears his name. Halley wanted to settle a dispute between his London friends Christopher Wren and Robert Hooke about the path a body would take if it was influenced by a force obeying an inverse-square law. He traveled to Cambridge in August 1684 and, in Newton's rooms, asked the great man. "Why, a body would travel in an ellipse," replied Newton immediately. "I have proved it."

But, try as he might, Newton was unable to find the proof among his mountain of papers. He promised Halley he would redo the calculation and send it to him in London. He was as good as his word. Halley was stunned by the nine-page pamphlet—"On the motion of bodies in orbit"—that arrived at his door several months later and urged Newton to publish it immediately; but Newton refused, saying he wanted to present all of his ideas on gravity and on motion in a book. Thus, he embarked on an eighteen-month gargantuan effort that culminated in the publication of the *Principia*, which rivals only Charles Darwin's *On the Origin of the Species* as the greatest book in scientific history.[4]

19.

THE PLANET THAT STALKED THE EARTH

Once upon a time,
the earth had a ring around it

"Which is more important: the sun or the moon?
The moon, obviously—the sun shines during the day
when it is light anyway."
—RUSSIAN RIDDLE

THE ORIGIN OF THE moon has been a longstanding puzzle. Of all the satellites in the solar system only the earth has a moon so close to it in size. At a quarter of the earth's diameter, it pretty much makes the earth-moon system a double planet.

The moons in the solar system appear to have two distinct origins. Some are merely pieces of space rubble that at some time wandered too close to a planet and were captured by its gravity. Others congealed out of a debris disk left over from a planet's birth in much the same way that the planets condensed from the leftover rubble swirling around the newborn sun. The moons that result from these two processes are a lot smaller compared with their parent planets than the moon is compared with the earth. This is one reason why planetary scientists have postulated an entirely different and pretty unique mechanism for the birth of the moon.

Imagine the earth shortly after its formation, 4.55 billion years ago. The solar system is a dangerous place. Huge space rocks—the building blocks of planets—are still careering around here, there, and everywhere. One of these *planetesimals*—a world the size of present-day Mars—is more dangerous than the rest. It is heading directly for the earth. When the impact comes, the collision is so violent that the entire exterior of the earth turns molten and is splashed off into space, forming a ring around the earth.

Many are convinced that this is how the moon came to be. The key evidence came from the Apollo astronauts. The moon rocks they brought back in their spacecraft revealed that the lunar material is suspiciously like the earth's mantle, and its rocks are drier than the driest terrestrial rocks, as if all their water was once driven out by intense heat. All these things are compatible with a mega-collision. The problem has been that a Mars-mass object would need to have dealt the earth a glancing blow at a very low velocity in order to create the moon and not in the process shatter the earth—yet bodies orbiting the sun either closer in or further out than the earth are moving far too fast for this to have been the case.

Nevertheless, the Big Splash theory, as it is called, can be made to work if the Mars-mass body—dubbed Theia—actually shared the same orbit as the earth. This could have happened if it formed at a stable *Lagrange point*, either sixty degrees behind or sixty degrees ahead of the earth in its orbit.[1] For millions of years, before the gravity of a passing body nudged into an Earth-colliding orbit, Theia bided its time. It was the planet that stalked the earth.

The ring around the earth that formed from the impact of Theia was short-lived. Its constituent debris cooled quickly then gradually congealed into a brand-new body: the moon. At first, the new satellite was ten times closer to our planet than it is today and raised tides on the earth 1,000 times higher than it does now. But raising those tides sapped energy from the earth-moon system.[2] It caused the earth's spin to slow down and the moon to recede until it reached its present location.

Energy continues to be sapped from the earth and moon. And, today, the moon is still moving away from the earth—at about 3.8 centimeters a year. In other words, during your lifetime the moon will move a car's length further away. We know this because we can bounce laser light off reflectors left on the lunar surface by American and Russian spacecraft. The fist-sized reflectors, known as "corner-cubes," have the property that they reflect back light in exactly the same direction it comes from. So, if the time of a flight of light from the earth to the moon is known, it is possible to deduce the distance of the moon by knowing the speed of light.

The corner-cubes left on the lunar surface fly in the face of the conspiracy theorists' claim that humans have never visited the moon. They were left by the manned American spacecraft Apollo 11, 14, and 15 and the unmanned Russian rovers *Lunokhod 1* and *2*.

The *Lunokhod 2* reflector works occasionally but the one on *Lunokhod 1* was lost for almost forty years. However, relatively recently its landing site was imaged by the Lunar Reconnaissance Observer. The coordinates were then passed to scientists in New Mexico. And, remarkably, on April 22, 2010, they fired a pulse of laser light at the landing site and were stunned to receive a return burst of 2,000 particles of light, or photons.

Our unusually big moon has been hugely important for life on Earth. The strong gravity of such a big moon stabilizes the earth's spin. If the planet tips over—as spinning tops tend to do—the moon pulls it upright again. Since such wobbles vary the sunlight reaching the ground, the moon stabilizes our climate. Mars, which has no big moon, suffers catastrophic climate change. Life on Earth could never have evolved without a stable climate over billions of years.

Our big moon also pulls large tides, which twice a day leave large tracts of the ocean margins high and dry. Long ago, stranded fish evolved lungs. Ultimately, this drove the colonization of the land.

Our big moon has even driven science. Total eclipses blot out the sun and make stars visible close to the solar disk. In 1919, this permitted the observation of the bending of starlight by the gravity of the sun, a key prediction of Einstein's theory of gravity. Isaac Asimov, in his 1972 essay "The Tragedy of the Moon," even claimed that had the moon belonged to Venus, rather than to the earth, science would have arisen 1,000 years earlier. Asimov argued that if people had seen Venus orbited by a visible moon, the geocentric idea of the earth as the center of creation would never have been viable—and the Church could not have silenced those who thought otherwise.

20.

PLEASE SQUEEZE ME

The body in the solar system that generates
the most heat, pound for pound, is not the sun

"I therefore concluded, and decided unhesitatingly, that there
are three stars in the heavens moving about Jupiter, as Venus and
Mercury about the sun; which at length was established as clear as
daylight by numerous other observations."

—GALILEO

THE BODY IN THE solar system that generates the most heat, pound for pound, is not the sun. It is Jupiter's giant pizza-like moon, Io.

On March 8, 1979, NASA's *Voyager 1* space probe was leaving Jupiter and heading off to a rendezvous with Saturn in late 1980. The Voyager team decided to take a parting shot of Io and so turned the camera back the way the space probe had come. The image it obtained was startling. Spouting from the tiny crescent moon, silhouetted against the starry backdrop of space, was a phosphorescent plume of gas.

Over the following days, the Voyager team spotted a total of eight gigantic plumes, pumping matter hundreds of kilometers into space. Io, it turns out, is the most geologically active body in the solar system, with more than four hundred rumbling volcanoes. Actually, the vents that pepper the orange and yellow and brown of its pizza-like surface are more reminiscent of the geysers of Yellowstone Park in the United States. And that is what they are: geysers, rather than volcanoes. Lava does not erupt directly from the moon's molten interior but instead superheats liquid sulphur dioxide just beneath the surface, converting it into gas that bursts from the vents exactly like the pressurized steam of a terrestrial geyser.

Every year, Io pumps about 10,000 million tonnes of matter into space. As it falls back in the moon's low gravity, it coats the surface with sulphur—just like the deposits around a Yellowstone fumarole. This is why the satellite has a pizza-like appearance. The lurid colors are the phases exhibited by the element sulphur at different temperatures.

The key to understanding Io's activity is the bodies in its neighborhood: Jupiter and the Galilean moons. Io is the innermost of the four Galilean moons, so-called because they were discovered by Galileo with his newfangled telescope in 1610. It orbits about as far from the giant planet as the moon is from Earth. But the enormous gravity of Jupiter—318 times as massive as the earth—whirls the moon around not in 27 days, like the earth's satellite, but in a mere 1.7 days.

Critical to Io's heating are two of the other Galilean moons that orbit further out from Jupiter than Io—Europa and Ganymede. Ganymede is actually the largest moon in the solar system, bigger even than the closest planet to the sun, Mercury. For every four circuits Io makes of Jupiter, Europa completes two and Ganymede one. Because of this, periodically the two satellites line up, reinforcing each other's gravitational tug on Io. The effect is to yank Io, elongating its orbit; so it swings in close to Jupiter and then flies back out again, repeatedly. And it is this motion, it turns out, that is behind Io's prodigious heating.

The difference in Jupiter's pull on the side of Io facing the giant planet and the side facing away causes the moon to bulge. When Io is at its closest in its orbit to Jupiter, this tidal bulge is bigger than when Io is at its most distant.[1] Up and down, up and down, the rock is stretched and squeezed. And, just like a rubber ball squeezed repeatedly in your hand, Io gets hot. In fact, its interior is heated to melting point.

We now know of hundreds of Jupiter-mass planets orbiting other stars. And there is every reason to believe they too are accompanied by giant moons, just like Jupiter's Io. Thanks to tides, such exomoons will have their own central heating. This means liquid water, the prerequisite of terrestrial biology, may be present on their surfaces even though they may be far from the warmth of their parent sun. Consequently, exomoons, rather than planets like the earth, may be the most likely locations in our galaxy to find extraterrestrial life.

21.

HEX APPEAL

At the north pole of Saturn is a hurricane
twice the size of the earth and the shape of a
hexagon

"Nature's imagination's so much greater than man's.
She's never gonna let us relax."
—RICHARD FEYNMAN

WHEN AIR CIRCULATES IN the earth's atmosphere, it circulates—well—in a circle. Have you ever seen a triangular hurricane? Or a square one? Or a hexagonal one? Of course you haven't. However, things are different at the north pole of Saturn.

In 2007 NASA's *Cassini* space probe flew over the ringed planet and captured the most bizarre image of a hexagonal arrangement of clouds turning around the pole. The hexagon is almost twice as wide as the earth. Peculiarly, however, there is no matching hexagon at the south pole of Saturn (poles on a gas planet ought to be similar), just clouds circulating around an eye, like clouds do around the continent of Antarctica on Earth.

Saturn's polar hexagon was first spotted by NASA's *Voyager 1* and *2* space probes when they flew by the planet a quarter of a century earlier. Evidently, the honeycomb-shaped weather system is very stable and long-lived.

A clue to the origin of the polar hexagon comes from laboratory experiments in which a fluid is spun rapidly in a bucket. Researchers have found that, under certain conditions, there spontaneously appears a pattern in the shape of a polygon with three, four, five, or six sides.[1] These unchanging *standing waves* are thought to be generated by the fluid bouncing off the walls of the bucket. The only problem when extrapolating to Saturn, of course, is that the atmosphere at Saturn's north pole is not in a bucket.

The cause of the hexagon, which rotates at roughly the same rate as Saturn, is now believed to be a jet stream, similar to the ones found on Earth but at least four times faster. A jet stream is a core of winds high in the atmosphere of a planet. No researchers have yet been able to model all the behavior of the hexagonal hurricane. However, a team led by Raúl Morales-Juberías of the New Mexico Institute of Mining and Technology claim to have created a model that best matches what we see.[2] The team simulated a jet circulating around Saturn's pole. When they *perturbed*, or jiggled the jet, it meandered into a hexagonal shape that rotated with very nearly the same period as Saturn's rotation.

Problem solved? Since nobody has yet come up with an explanation that has received universal acceptance, we'll have to wait and see.

22.

MAP OF THE INVISIBLE

The planet Uranus was originally named...
George

"I have looked further into space than ever
human being did before me."
—WILLIAM HERSCHEL

URANUS WAS DISCOVERED IN 1781. It was discovered from a back garden in the English town of Bath. And it was discovered by a German freelance musician called William Herschel.

At nineteen, Herschel had moved from Hanover with his sister, Caroline, to be the organist at a church in the spa town, famous since Roman times for its hot springs. But, although music was the means by which Herschel earned a living, astronomy was his true passion. And in Bath he constructed some of the most powerful telescopes of his day. It was with one of these instruments, on the night of March 13, 1781, that he spotted a fuzzy star, which he at first thought was a comet. However, over the next few nights, as it crept across the backdrop of fixed stars, he noticed that it did not follow the highly elongated orbit of a comet but instead the near-circular orbit of a planet.

Herschel had found the first planet unknown to the ancients—the first world to be discovered in the age of the telescope. It orbited in the chilly darkness far beyond Saturn, long believed to be the outermost planet. Overnight, he had doubled the size of the solar system.

Herschel was keenly aware of his immigrant status and had a great desire to be accepted by his adopted country. He therefore named the new planet "George," after the English king, George III.[1] Had the name stuck, then today we would have, in order of increasing distance from the sun, Mercury, Venus, Earth, Mars, Jupiter, Saturn…and George!

The French vehemently objected to having a planet named after an English king and instead referred to it as "Herschel." It was the German astronomer Johann Bode who suggested that the planet be called Uranus, after the father of the Roman god Saturn. The name stuck.

It turned out that Uranus had been spotted almost a century earlier by English astronomer John Flamsteed. In 1690, mistakenly believing it to be a star not a planet, he had catalogued it as 34 Tauri, the thirty-fourth star in the constellation of Taurus.

Because of the earlier sightings of Uranus, it was possible quite early on to trace its path across the heavens and therefore deduce its orbit around the sun. But, by the mid-nineteenth century, it was obvious that something was wrong. Uranus was not following the elliptical path dictated by Newton's law of gravity. Predictions of where it would be at any future date always proved wrong.

Enter French astronomer Urban Le Verrier. He guessed that, even further from the sun than Uranus, there must be an unknown massive planet whose gravity was tugging at Uranus and perturbing its orbit. The calculations required to pin down the location of the hypothetical planet were horrendously complicated but, after a great struggle, Le Verrier succeeded with them. The only problem was that he could not persuade the director of the Paris Observatory to direct his astronomers to look for the planet. In desperation, on September 18, 1846, he wrote to Johann Galle at the Berlin Observatory.

Galle had, the previous year, sent Le Verrier his thesis—and Le Verrier had failed to even acknowledge its receipt. Fortunately, Galle bore no grudges. The director of the Berlin Observatory, Johann Encke, was hardly more enthusiastic than his French counterpart to give up telescope time for what he saw as wild-goose chase. However, because Encke would be celebrating his fifty-fifth birthday on the night of the September 23 and therefore doing no astronomy, he gave Galle permission to use the observatory's twenty-two-centimeter Fraunhofer telescope.

On the morning of September 24, 1846, within an hour of beginning the search, Galle and astronomy student Heinrich d'Arrest found a new planet exactly where Le Verrier had said it would be. This was a stunning moment in the history of science: it was now possible to predict the existence of things never before suspected. Newton's theory of gravity not only explained what astronomers could see in the night sky but also revealed what they could *not* see. It provided a "map of the invisible."

The new planet was named Neptune. Its discovery created a sensation and made a superstar of Le Verrier.[2] It even set him on a true wild-goose chase—searching for a hypothetical planet orbiting closer to the sun than Mercury and dubbed "Vulcan."[3]

Newton's theory of gravity has proved to be the gift that keeps on giving. Dark matter is the Neptune of today. We know it outweighs ordinary matter by a factor of six because we can see it tugging with its gravity on the visible stars and galaxies. But, as yet, we have no idea what it is.

23

LORD OF THE RINGS

Galileo thought Saturn was a planet with...ears

"The scientific theory I like best is that the rings of Saturn
are composed entirely of lost airline luggage."
—MARK RUSSELL

GALILEO GALILEI WAS A giant in the history of science. Among other things, he discovered that the swing of a pendulum is perfectly regular and that all bodies, regardless of their mass, fall at exactly the same rate under gravity. But arguably a low point in his career came in 1610 when he pointed his newfangled telescope at Saturn and declared it to be "a planet with ears." In 1611, he changed his mind and speculated that the planet had a moon on either side, each a third as big as the planet. But, in 1612, to his consternation, the two moons vanished. "Saturn has swallowed its children!?" he wrote to his patron, the Grand Duke of Tuscany. In 1613, the moons appeared again, and Galileo was even more baffled.

Unfortunately, Galileo was destined to die never having made sense of his observations of Saturn. The telescope he aimed at the night sky from Padua was simply not powerful enough to reveal the planet's big secret. The mystery was solved only half a century later, in 1655, when the Dutch scientist Christiaan Huygens built an improved telescope with a magnification of fifty. Huygens recognized, correctly, that Saturn is girdled by a wide system of rings.

Nowadays, we know that the plane of Saturn's rings is tilted at 26.7 degrees to our line of sight. The rings maintain their orientation in space, much like a spinning gyroscope, but, as Saturn orbits the sun, the earth is presented with the rings at various different angles. When we see the rings edge-on—twice during Saturn's twenty-nine-and-a-half-year orbit—they vanish totally from view. And, at other times, when the rings appear at an angle to the line of sight, they do indeed look like ears.

Three other planets in the solar system have rings—Jupiter, Uranus, and Neptune—but none of the other ring systems are anywhere near as impressive as those of Saturn. Its rings extend almost from the planet's cloud tops to 140,000 kilometers from the center of the planet. To give you some idea what this means, if they girdled the earth they would stretch more than a third of the way to the moon.

As soon as Saturn's rings were discovered, the biggest question was: what are they made of? This matter was addressed in the nineteenth century by James Clerk Maxwell, the greatest physicist between the time of Newton and Einstein, whose supreme achievement was to distil all electrical and magnetic phenomena into one set of equations, discovering in the process that light is actually a wave of electromagnetism.[1] In 1858, in a tour de force of mathematics, Maxwell showed that, if the rings are either a solid or a fluid, they would disintegrate. He concluded that they must be made of countless particles sweeping around Saturn—a swarm of micro-moons, if you like.

The definitive proof of this came more than a century later when NASA's *Voyager 1* and *Voyager 2* space probes flew past Saturn in 1980 and 1981. Although from Earth astronomers see only a handful of wide rings interrupted by gaps such as the "Cassini Division,"

Voyager's cameras revealed tens of thousands of narrow ringlets. The inner ringlets orbit Saturn faster than the outer ones, confirming Maxwell's conclusion that they could never be solid.

The rings are actually made of countless shards of 99-percent-water ice, sparkling in the sunlight, which is why they appear so spectacularly bright. Their sizes range from smaller than a sand grain to an office block. The brightest rings must be made of particles with a large reflective area—possibly fluffy, snowflake-like agglomerations, which constantly come together and fall apart. All the ring particles orbit in a layer probably less than twenty meters thick. To put it another way, if the rings were shrunk to a disk a kilometer across, they would be thinner than the sharpest razorblade.

If you put all the ring particles together, they would form a body between about 200 and 300 kilometers across. This is the size of a medium-sized moon of Saturn and may be a clue to the origin of the rings. Conceivably, a moon drifted too close to Saturn and was torn apart by its gravity. Or perhaps a moon was shattered when it was struck by a comet or an asteroid.

The fact that the rings are so bright indicates that they must be less than about four hundred million years old, since dust from meteorite impacts is expected to darken them over time. However, because four hundred million years is less than a tenth the age of the earth, it would mean that we are very lucky to be around to see the rings. Scientists hate invoking luck as an explanation for anything. One way the rings can be old but appear young is if the ring material is continually clumping together and being shattered by meteorite impacts. This recycling would be akin to continually breaking open snowballs to reveal pristine ice inside.

In a final twist to the story of the rings, it turns out that Saturn does not actually have rings at all—it has multiple spirals, like the grooves on an old vinyl record. Vibrations of the icy rubble—perhaps created by meteorite impacts—cause a *spiral density wave* to move outwards. As it passes, it compresses together particles, making temporary ringlets. Spiral density waves also create the spiral arms of our Milky Way. Amazingly, Saturn's rings are just a tightly wound version of our spiral galaxy.

24.

STARGATE MOON

On a moon of Saturn is a mountain range
twice the height of Everest that was built...
in an afternoon

"My God, it's full of stars!"
—DAVE BOWMAN, *2001: A SPACE ODYSSEY*[1]

APETUS IS THE SITE of the "stargate"—the portal to another part of the galaxy—in the novel *2001: A Space Odyssey*. This icy moon of Saturn was selected by science-fiction writer Arthur C. Clarke because, mysteriously, one face is about ten times brighter than the other. What better place to locate an artificial alien artifact, reasoned Clarke, than on a moon that itself appears to be artificial?

Why Iapetus is Janus-faced is one of the longest-standing mysteries in astronomy. It has persisted since the moon's discovery by the astronomer Giovanni Domenico Cassini in 1671. The probable explanation, however, is an encounter not with an enigmatic extraterrestrial race but with Saturn's spectacular system of rings.

The key to unravelling the mystery was the flyby of Iapetus by NASA's *Cassini* space probe on December 31, 2004. Its camera captured images of the crater-strewn moon in unprecedented detail. And they revealed a feature that left planetary scientists amazed.

Stretching for 1,300 kilometers—almost a third of the way around the moon—is an extraordinary mountain ridge. In places it is twenty kilometers high, which is more than twice the height of Mount Everest. And this is on a moon only 1,436 kilometers across, much less than half the diameter of our own moon. The ridge follows the equator closely. And there is nothing else like it anywhere else in the solar system.

Although the ridge appears to be a yet another puzzle to add to the mystery of the two-tone coloring of Iapetus, perhaps significantly it perfectly bisects the dark region of Iapetus. Might the two features be linked? This was certainly the suspicion of planetary scientist Carolyn Porco, head of the *Cassini*'s imaging team.

According to astronomer Paulo Freire of the Arecibo Observatory in Puerto Rico, Saturn's rings might provide the link. Once upon a time, he suggests, Iapetus wandered too close to Saturn's ring system. The rings are possibly the fossil relic of one or more shattered moons. Although they are only about twenty meters thick, they are crammed with pieces of tumbling ice ranging in size from a dust grain to a terrace of houses. When Iapetus blundered into the outer edge of the rings, it was like a rock meeting a strimmer. All along the line where the rings intercepted the moon, the surface came under ferocious bombardment.[2]

According to Freire, the bombardment deposited an enormous volume of material in a relatively short time. He calculated the volume. First, he assumed that Iapetus and Saturn's ring system were in contact for just hours, a small fraction of the seventy-nine days it takes the moon to orbit Saturn. From the density of material in Saturn's rings and the relative speed of the moon and rings—about ten times that of a jumbo jet—Freire concluded that an astonishing twenty-five million cubic meters of ring material would be dumped on every meter of ridge line. Enough to create a ridge five kilometers high with a base width of ten kilometers.

Had Iapetus penetrated the ring system completely, the resulting ridge would have extended not just a third of the way but halfway around the moon. From this, Freire concluded that the moon merely grazed the rings.

But even if this scenario is correct, and a collision with Saturn's rings created a mountain range twice the height of Everest in an afternoon, there remains the puzzle of why the face of Iapetus centerd on the ridge is so black. Although Saturn's rings are at least 99 percent pristine ice, which is why they shine so brightly, they also contain a perfect source of blackening material—tiny dust grains. So, how did this dust get transported from the ridge to cover almost a complete hemisphere of the moon? Freire suggests the key is the composition of the ice that makes up Iapetus. Some is carbon dioxide, or "dry" ice, which is also present in the ring material. In the violence of the collision, this material would have turned into vapour, creating a temporary atmosphere on the tiny moon.

Carbon dioxide actually *sublimates*, going directly from a solid to a vapour in an explosive manner. Consequently, within the temporary atmosphere, thin but super-fast winds spread out from the ridge, depositing dust over a large area of Iapetus. This should cause the amount of dust—and the blackening of Iapetus—to drop gradually with distance from the ridge. And this is exactly what is observed.

The only way Iapetus could have hit Saturn's ring system is if it orbited in the same plane as the rings. The ridge would then have been created along its equator. However, today, Iapetus no longer orbits in the plane of the rings. So, something must have knocked it into its present orbit; most likely a collision with another moon.

This might have been possible if there had once been many moons in similarly chaotic orbits around Saturn. It is not an unlikely scenario. The rocky planets of the solar system were built up by the collision and sticking together of many smaller objects in chaotic orbits around the sun. And the moons of Saturn were similarly built up by the collision of many small objects around Saturn. In the case of Saturn, all were shattered, incorporated into other moons or ejected from Saturn. All, that is, except a relatively large one—Iapetus.

Freire, however, is not the only scientist to have proposed an explanation of the black-and-white coloring and gargantuan ridge on Iapetus. Other theorists have suggested that in the distant past an icy body hit Iapetus, ejecting material that formed a moon—a moon around a moon. Gradually, this mini-moon spiraled in towards Iapetus, eventually shattering into myriad shards that were then deposited on the surface to form the titanic ridge.[3] Still other theorists favor an explanation in which Iapetus was born spinning very fast, causing material to fly outwards and create an equatorial ridge.[4] Sadly, however, none of the competing theories postulates the existence of a stargate!

PART FIVE

FUNDAMENTAL THINGS

25

INFINITY IN THE PALM OF YOUR HAND

You could fit the entire human race in the
volume of a sugar cube

"To see a World in a Grain of Sand
And a Heaven in a Wild Flower
Hold Infinity in the palm of your hand
And Eternity in an hour"
—WILLIAM BLAKE, "AUGURIES OF INNOCENCE"

YOU COULD FIT THE entire human race in the volume of a sugar cube. The reason for this is that matter is mind-bogglingly empty. You probably have a picture in your mind from school of an atom, the fundamental building block of all matter. Most likely, it is an image of a miniature solar system with a nucleus sitting at the center like a sun around which electrons orbit like planets. However, this image fails to convey just how much empty space there is inside an atom. The playwright Tom Stoppard said it best: "Now make a fist, and, if your fist is as big as the nucleus of an atom, then the atom is as big as St Paul's, and if it happens to be a hydrogen atom then it has a single electron flitting about like a moth in an empty cathedral, now by the dome, now by the altar."

As a percentage, the amount of empty space in an atom is about 99.9999999999999. You are a ghost. I am a ghost. We are all ghosts.

And, if you could squeeze all the empty space out of all the seven billion people in the world, the human race would indeed fit in the volume of a sugar cube (albeit a very heavy sugar cube!).

This is not mere theoretical fantasy. There are objects out in the universe whose atoms have had all the empty space squeezed out of them. They are called neutron stars and they are the endpoint of the evolution of very massive stars. When such a star blows its outer layers into space as a supernova, its core, paradoxically, implodes (in fact, it is the implosion that is believed to drive the explosion). The resulting neutron star is about the size of Mount Everest but contains about the mass of the sun. And if you could travel to a neutron star and dig out a chunk the size of a sugar cube, you would indeed find it weighed as much as the human race.

But why are atoms so empty? The answer is provided by quantum theory. Quantum theory is our very best description of the microscopic world of atoms and their constituents. It is fantastically successful. It has given us lasers and computers and nuclear reactors. It explains why the sun shines and why the ground beneath our feet is solid. In fact, it is the most successful physical theory ever devised, predicting what we observe in experiments to an obscene number of decimal places. But, in addition to being a fantastic recipe for building things and for predicting things, quantum theory provides a unique window on an Alice in Wonderland world that exists beneath the skin of reality. It is a place where a single atom can be in two places at once (the equivalent of you being in London and New York at the same time); where things happen for absolutely no reason at all; and where two atoms can influence each other instantaneously even if on opposite sides of the universe.

All this quantum insanity stems from a simple observational fact: the fundamental building blocks of matter have a weird dual nature.

They can behave as localized particles like tiny billiard balls, and they can behave as spread-out waves like ripples on a pond.[1] Don't even try to imagine how this can be. It is impossible. The truth is that the electrons and photons and so on that make up the world are neither particles nor waves but something for which we have nothing to compare them with in the everyday world and for which we have no word in our vocabulary. As with an object we can never see directly and can know only from the shadow that it casts on two adjacent walls, we can never actually see the denizens of the quantum world, only the shadows they cast in our experiments: one of a tiny bullet, the other of a dancing ripple.

The key thing is that the smaller the particle, the bigger its quantum wave.[2] The smallest particle of familiar matter is the electron; it therefore has the biggest quantum wave. And it is because the electron wave needs tons of elbow room that atoms have to be so big relative to their nuclei—why they contain so much empty space.[3]

In fact, the wave nature of the electron is why atoms exist. As the American Nobel Prize-winning physicist Richard Feynman said: "Atoms are completely impossible from the classical point of view." What Feynman meant is that, according to the theory of electromagnetism, an electron whirling around a nucleus in an atom should constantly broadcast "electromagnetic waves" like a tiny radio transmitter. This should cause it to lose energy and spiral into the nucleus in less than a hundred-millionth of second. Atoms should collapse in on themselves. They should not exist.

This rather contradicts reality, however, where atoms have existed for at least the age of the universe—13.82 billion years. This is about one followed by forty zeroes longer than predicted by the theory of electromagnetism![4]

Quantum theory comes to the rescue because electron waves are spread out and always take up a minimum amount of space. They cannot be squeezed down into a nucleus. Thus, atoms—thank goodness, since we are all made of atoms—are able to exist.

If there is one thing that perfectly encapsulates the madness of the quantum world it is this: the British physicist J. J. Thomson won the Nobel Prize for showing that the electron is a particle. His son, George Thomson, won the Nobel Prize for showing that it isn't. I imagine Thomson family get-togethers being raucous affairs, with J. J. shouting, "It's a particle!" and his son shouting back, "No, it's not!"

26.

BUNGALOW BENEFITS

You age faster on the top floor of a building than on the ground floor

> "I can't talk to you in terms of time—
> your time and my time are different."
> —GRAHAM GREENE

YOU AGE FASTER ON the top floor of a building than on the ground floor. This is a consequence of Einstein's theory of gravity—the general theory of relativity—which dictates that time flows more slowly in strong gravity. Since the ground floor of a building is closer than higher stories to the mass of the earth, gravity is marginally stronger there, and so time flows marginally more slowly. (So, if you want to live longer, move to a bungalow!)[1]

The slowing of time by gravity is an extremely tiny effect, and detecting it requires a super-sensitive atomic clock. Incredibly, in 2010, physicists at the National Institute of Standards and Technology in the United States were able to show that, if you stand on one step of a staircase, you age faster than a person standing one step below you. They did this by using two super-accurate atomic clocks, one on each step.[2]

So why do you age more slowly in strong gravity? Well, it was Einstein's genius in 1915 to realize that the force of gravity doesn't actually exist. That's right, it is an illusion! This takes a bit of getting your head around. But imagine you wake up in a rocket in empty space, which is accelerating at 1g. You find that your feet are glued to the floor, and you are able to walk around exactly as if you are on the surface of the earth. In fact, if the windows are blacked out and you do not know you are in a rocket, you may actually conclude you are in a cabin on the surface of the earth. According to Einstein, this reveals an amazing truth about gravity: gravity is acceleration. At this moment, you may think you are glued to the ground by a force, but actually you are merely accelerating and do not know it.

Wait a minute. How can this be?

Imagine the rocket again. Say you shine a laser beam horizontally from the left side of the rocket to the right side. If you look closely, you will see that the beam strikes the right wall at a point slightly closer to the floor than when it started out. This is because, in the time the light has been flying across the cabin, the floor has accelerated upwards at 1g towards the light beam. But, of course, you do not know you are accelerating in a rocket. You think you are experiencing gravity in a cabin on Earth. How then do you explain the fact that the light beam appears to have curved downwards towards the floor?

A well-known property of light is that it always takes the shortest path between two points. For a hiker walking across flat terrain, the shortest path between two points is a straight line. However, this is not the case for a hiker walking through a hilly landscape. The shortest path is wiggly (think of the path from the point of view of a high-flying bird). From the fact that the laser beam follows a curved path across the cabin you can only conclude that space is warped, or curved, like the hilly landscape. And, since you believe you are experiencing gravity, this means that gravity curves space. In fact, as Einstein realized, gravity *is* curved space.

Actually, gravity is curved space-time, because, as Einstein discovered earlier, in 1905, space and time are actually aspects of the same thing. We do not realize that space-time is curved because space-time is a four-dimensional thing and we are lowly three-dimensional beings. That is why it took a genius like Einstein to recognize the truth.

So the true explanation of why, on the surface of the earth, our feet are glued to the ground is that the space-time around the earth is curved. It is as if the planet is at the bottom of a valley of space-time. We are accelerating down to the bottom of the valley, but our way is blocked by the surface of the earth. It is the surface pushing upwards on our feet that we interpret as gravity.

Since gravity is warped space-time, it follows that not only does it play games with space—bending the paths of light beams—but it also plays havoc with time. And here—at last!—we come to the explanation of why time is slowed by gravity.

Picture an idealized "clock" that consists of a horizontal laser beam bouncing back and forth between two mirrors. Each time the light strikes a mirror, it is detected, creating a "tick." If the clock is on the earth, then the light does not travel between the mirrors in a perfectly straight line but follows a curved path—because, of course, gravity bends light.

Now think of two such clocks—one higher above the ground than the other. The lower clock is in slightly stronger gravity than the higher clock since it is closer to the bulk of the earth. This means that the light traveling between the mirrors in the lower clock follows a more curved path than the light in the higher clock. The more curved a path, the further the light has to travel between the mirrors, and the greater the time between ticks. It follows that the lower clock ticks more slowly than the upper clock. In other words, time flows more slowly in strong gravity.

What I have described is the most basic clocks imaginable. If gravity affects this clock, it affects every clock. There is no way to evade the conclusion that gravity slows time.

Now, if you think this is an esoteric effect with no relevance to everyday life, think again. SatNavs and smartphones determine your location with respect to a constellation of GPS satellites, which swing round the earth in highly elongated orbits. GPS satellites carry on-board clocks, and when the satellites swoop in close to the planet, they experience stronger gravity and their clocks slow down. If your electronic devices did not compensate for this effect, they would be unable to pinpoint your location relative to the GPS satellites.

In other words, many of us on a daily basis are inadvertently taking part in an experiment that tests Einstein's general theory of relativity. Were the theory false, then the GPS system would get your location wrong by about an extra fifty meters a day. In fact, after ten years, it will still be correct within five meters, showing just how accurate general relativity is.[3]

The slowing down of time by gravity is small where gravity is weak but significant in places where gravity is strong. And the strongest source of gravity we know of is a black hole, a bottomless pit in space-time left by the catastrophic shrinkage of a massive star at the end of its life. If you could hover just outside a black hole's *event horizon*—the point of no return for in-falling light and matter—time would slow so much relative to the rest of the cosmos that if you looked outwards, you would see the entire future history of the universe flash past your eyes like a movie in fast-forward!

27.

THE INCREDIBLE EXPLODING MOSQUITO

A force of unimaginable power
is operating in each and every one of us—
but we never notice it

> "Ten thousand engines in ten thousand places running the
> machines of industries and homes—all turning because of
> the knowledge of electromagnetism."
>
> —RICHARD FEYNMAN[1]

I AM HOLDING AN EMPTY jam jar. I take it to public talks where I hold it up and say: "I've brought along my pet mosquito, Terry. Can you all see him? He's quite small." Some people actually come up to me afterwards and ask: "Is there really a pet mosquito in that jar?" Others, a bit more tongue-in-cheek, express concern about Terry's welfare: "Are you sure there are enough holes in the lid for him to breathe?"

When I hold up my jam jar, I tell the audience that Terry, like all of us, is made of atoms. And an atom, as we learn at school, consists of a nucleus, like a sun, about which electrons orbit like planets. The nucleus has a positive electric charge and the electrons a negative electric charge. "Unlike" charges have the property that they attract each other, and it is this attractive force that glues together an atom.

Next, I ask the audience what would happen if, as by magic, it became possible to remove all the electrons in Terry? That would leave only the positively charged nuclei—and "like" charges have the property that they repel each other. Terry would therefore explode. The question is: with what energy would Terry explode? The energy of...

1. a sparkler
2. a stick of dynamite
3. an H-bomb
4. a global mass extinction

Very few audience members plump for a sparkler or a stick of dynamite. Most of them, perhaps because they sense a trick question, stick up their hands for an H-bomb. But, actually, the answer is 4) —Terry would explode with an energy equivalent to a global mass extinction—that is, equivalent to the energy of the city-sized asteroid that slammed into the earth sixty-six million years ago and wiped out the dinosaurs.

The whole point of this is to emphasize the mind-boggling strength of the electromagnetic force that glues together the atoms in our bodies. Gravity appears strong. After all, you can jump barely a meter in the air before it drags you back down to the earth. But the electromagnetic force is stronger. Not by a factor of ten. Or a factor of one hundred. Or even a million. The electromagnetic force is stronger than the force of gravity by an astonishing factor of 10,000 billion billion billion billion.[2]

So, why, when you walk past someone on the street, do neither of you feel this phenomenally powerful force? Well, think of gravity. As far as this force is concerned, there is only one type of matter: that which always attracts. However, for the electromagnetic force, there are two types of matter: that which enables the force to attract and that which enables it to repel. By convention, we call the two types positive electric charge and negative electric charge. And it turns out that all normal matter is made of exactly the same amount of positive charge as negative charge, which means that the attraction and repulsion is perfectly balanced. So, in all everyday circumstances, the electromagnetic force is completely nullified.

This is why nobody guessed that such an incredibly powerful force existed until relatively recently in history—although there were hints. Think of lightning bolts. These occur when powerful up-drafts of air in a storm separate electric charge. The mechanism is still not well understood; but, eventually, the separation of charge between a cloud and the ground becomes so great that the force between separated charges breaks apart the very atoms of air, unleashing a surge of electrons to the ground that brings everything back into balance again.

In the past century or so, we have learned how to create artificial charge imbalances and how to unleash the mind-bogglingly powerful electromagnetic force. In fact, in a nutshell, the creation of charge imbalances is the secret of how electricity powers the modern world.

But why is electromagnetic force 10,000 billion billion billion billion times stronger than the force of gravity? That is the $64,000 question. We know of four fundamental forces that glue together the particles of matter, and we have a strong suspicion they are all merely aspects of a single *super force*. But it is hard to see how a single theory—that is, a single equation—could give rise to forces that differ so spectacularly in strength. Why the electromagnetic force is so wildly different from the gravitational force remains one of the outstanding questions in physics. As physicists attempt to answer it, they stand at the absolute frontier of the known, staring out into the fog of the unknown.

28.

THE UNKNOWABLE

Most things cannot be computed
by computers

"Computers are useless.
They can only give you answers."
—PABLO PICASSO

A MICROWAVE OVEN IS ALWAYS a microwave oven. It can never be a vacuum cleaner or a toaster or a nuclear reactor. However, a computer can be a word processor or an interactive video game or a smartphone. The list is endless. This illustrates the unique feature of a computer: it can simulate any other machine.

But what are the limits of what computers can do? Enter Alan Turing, the English mathematician famous for his role in breaking the Nazi Enigma and Fish codes and, arguably, shortening the Second World War by several years. In the 1930s, before any practical computers existed, Turing asked: "What are the limits of computers?" The answer he found was very surprising.

At its most fundamental level, a computer is a shuffler of symbols. A bunch of symbols goes in—perhaps the height, speed, and so on, of an airplane—and another bunch of symbols comes out—for instance, the amount of jet fuel to burn, the necessary changes to be made in the angle of ailerons, and so on. What transforms the input symbols into the output symbols is a set of instructions stored internally. Crucially, this program is infinitely rewritable. This is the reason that a computer can simulate any other machine. This is the source of its extraordinary versatility.

Turing, on a piece of paper, imagined an abstract machine that shuffles symbols on the basis of a stored program. Its program is stored on a one-dimensional tape as a series of 0s and 1s—because everything, including numbers and instructions, can be reduced ultimately to such a binary code. Precisely how it works—a read/write head changing the digits one at a time—is not really important. The crucial thing is that Turing's machine can be fed a description, encoded in binary, of any other machine and will then simulate that machine.

Because of this unprecedented ability, Turing called it a Universal Machine. Today, it is referred to as a Universal Turing Machine. Clearly, it is unrecognizable as a computer. But that is exactly what it is. A Universal Turing Machine is the simplest computer imaginable: the irreducible atom of computing.

Ironically, Turing devised his machine-of-the-mind to show not what a computer can do but what it *cannot* do. As a mathematician, it was the ultimate limit of computers that interested him.

Turing's search for an incomputable problem did not take him long. In fact, hardly had he begun looking when he stumbled on a simple task that no computer, no matter how powerful, could ever carry out. The task is this: if a computer is given a program, can it tell whether the program will get caught in an endless loop, going around and round the same set of instructions forever like a demented hamster in a wheel, or whether it will eventually halt?

At first sight, Turing's "halting problem" seems utterly trivial. Surely, to see whether a program chugs through its calculations and comes to a halt or not, it is necessary only to run the computer program? Yes, but what if it gets caught in an endless loop only after a year? Or a century? Or a billion years? Now perhaps you begin to see the problem. The only way to be sure the program will eventually halt is to determine this in advance of running the program. So, is this possible? The answer, Turing discovered, is: no!

In 1936, by clever logical reasoning, Turing showed that deciding whether a program eventually halts or goes on forever is impossible and therefore beyond the capability of any conceivable computer. It is "incomputable."[1]

The fact that it was so easy to find a simple thing that no computer, no matter how powerful, could ever do is a hint that all is not rosy in the garden of computation. Incredibly, it turns out that most problems cannot be computed by computers. It's as if there exists a small archipelago of computable problems—which mathematicians have found—lost in a vast ocean of incomputable problems.

Thankfully, the halting problem turns out not to be typical of the kind of problems we use computers to solve—run spreadsheets, operate smartphones, fly passenger planes, and so on. Turing's limit on computers has, therefore, not held us back technologically. And, despite the rather surprising fact that computers were actually born in the abstract field of pure mathematics as machines of the imagination, they have turned out to be immensely practical devices.

A footnote to all this is that there is a deep connection between Turing's discovery of in-computability and a shocking discovery made in mathematics only five years earlier by the Austrian logician Kurt Gödel.

Each field of mathematics, it turns out, has the same simple structure. Mathematical truths, or theorems, are obtained from a series of self-evident truths, or axioms, by applying the rules of logic. In 1900, the German mathematician David Hilbert pointed out that finding theorems from axioms by applying logic was a mechanical process. It did not require the intuition or flair of mathematicians. In theory, all of mathematics could be obtained from a small number of axioms by blindly cranking through the rules of logic. Hilbert, in describing this mechanical process—although he did not know it—was describing the actions of a computer.

Enter Gödel. In 1931, he dashed all of Hilbert's hopes. He showed that there were theorems that could never be proved either true or false. They were *undecidable*. This following picture may help. Think of axioms as foundation stones and logic as scaffolding that connects them to theorems, which float high in the air like balloons! According to Gödel, there will always be free-floating balloons that can never be reached by scaffolding from the foundations. Of course, it is always possible to add more foundation stones. But that would create new unreachable balloons.

Gödel's undecidability theorem—more commonly known as his incompleteness theorem—is one of the most famous and shocking results in the history of mathematics. On its publication, it so depressed mathematicians that many gave up their profession in despair. And who could blame them? For, just as most problems are not computable, most mathematical statements are not decidable. There is an archipelago of provable mathematical truths lost in a vast ocean of unprovable truths.

All is not lost, however. Just as the halting problem turned out to be not typical of the kind of problems for which we use computers, the particular theorem that Gödel demonstrated was undecidable turned out not to be typical of the kind of theorems mathematicians are good at finding. Mathematicians kept their jobs.

The question is: how are mathematicians able to find provable truth in such a vast ocean of unprovable ones? By definition, each island in the archipelago is unreachable from every other island by applying the rules of logic. Perhaps this is an indication that, as some claim, the human brain is more than a computer and can do things that no conceivable computer can ever do.

29.

DOUBLE TROUBLE

An atom can be in two places at once—
the equivalent of you being in London
and New York simultaneously

"The universe is not only queerer than we suppose,
but queerer than we can suppose."
—J. B. S. HALDANE

AT THE BEGINNING OF the twentieth century it was discovered that the world of atoms and their constituents was nothing at all like the everyday world. With hindsight, perhaps nobody should have been in the least surprised. After all, it would take ten million atoms laid end-to-end to space the full stop at the end of this sentence. Why would the realm of the ultra-small dance to the tune of the same laws as tables and chairs and people?

But physicists did not simply discover that the microscopic world behaves differently; they discovered that it behaves in a way so ridiculously crazy that it appears impossible. "I remember discussions which went through many hours until very late at night and ended almost in despair," said the German physicist Werner Heisenberg. "And when at the end of the discussion I went alone for a walk in the neighboring park I repeated to myself again and again the question: can nature possibly be so absurd as it seemed to us in these atomic experiments?"

What we now know is that the ultimate building blocks of matter such as atoms, electrons, and photons have a peculiar dual nature. They can behave both as localized particles—like tiny billiard balls— and as spread-out waves—like ripples on the surface of the pond. Nothing in the everyday world behaves in this manner, so do not worry if you cannot get your head around it. Nobody can.

Quantum waves are decidedly odd. Rather than being physical waves, like ripples on water, they are abstract mathematical waves whose spread through space is determined by the Schrödinger equation. Where the quantum wave is big—that is, it has a large up-and-down motion, or *amplitude*—there is a high chance, or probability, of finding a particle; and, where the wave is small, there is a small chance of finding a particle.[1]

One obvious consequence of the *wave-particle duality* of nature's fundamental building blocks is they can do all the things waves can do. Waves can bend around corners. If they could not, you would not be able to hear a car backfiring in a neighboring street. But there is one thing that waves can do that is unremarkable in the everyday world but in the microscopic realm has utterly earth-shattering consequences.

Imagine there is a storm at sea, which creates giant rolling waves. The next day, when the storm has passed, the breeze merely ruffles the sea, creating tiny ripples. Now, anyone who has witnessed such waves will know that it is possible to have a combination of them: big rolling waves whose surface is rippled. And this turns out to be a general property of waves: if two or more waves are possible, so too is a combination of those waves.

Such a *superposition* has an extraordinary consequence in the microscopic world. Imagine a quantum wave representing an oxygen atom in the air, which is very large on the left-hand side of a room, so there is an almost 100 percent chance of finding the oxygen atom there. Now, imagine there is a quantum wave representing the same oxygen atom, which is very large on the right-hand side of the room, so there is an almost 100 percent chance of finding the atom there. Since both waves are possible, a superposition of the two waves is also possible. But this corresponds to the oxygen atom being on the left-hand side of the room and the right-hand side at the same time—to the atom being in two places at once.

Nature permits its basic building blocks to behave in this bizarre way but, paradoxically, we never actually observe it. If the oxygen atom is found to be on one side of the room, the part of the superposition representing the particle on the other side instantly disappears, or collapses, in a manner that is not completely understood. Basically, doing an experiment to reveal the localized, particle-like face of a quantum entity automatically precludes it revealing its non-localized, wave-like face.

But, if nature permits an atom to be in two places at once but we can never observe it, surely this is of no importance? Well, actually, no. Because it has consequences. Consequences that lead to all sorts of quantum weirdness. And, actually, to the very existence of the world we live in.

It is all down to another wave phenomenon, known as *interference*. We have all seen raindrops splashing in a puddle. Concentric waves spread out from each impact and overlap each other. Where two crests coincide, the waves reinforce; and where a crest of one wave coincides with the trough of another, they cancel out. If you were to insert a straight vertical barrier where the waves overlap, you would see locations where the wave was boosted alternating with locations where the water was calm. In fact, this experiment was done by English polymath Thomas Young in 1801. He engineered a situation in which two sources of spreading light overlapped and inserted a white screen. Lo and behold, on the screen, he saw alternating light and dark stripes not unlike a modern-day supermarket barcode. By demonstrating that light displays a wave phenomenon—interference—he proved that light is a wave. This is not at all obvious. The reason is that the wavelength of light—the distance between successive wave crests—is about a thousandth of a millimeter and so far too small to be discerned with the naked eye.

So, if an atom being in two places at once has consequences because of interference, what are they?

Imagine rolling together two bowling balls so that they ricochet off each other and fly away in opposite directions.[2] If you picture a clock face, they might for instance fly off in the direction of one o'clock and seven o'clock. Or nine o'clock and three o'clock. Now, if you collide the balls thousands of times, you will discover that they fly to every number on the clock face.

But now imagine colliding two microscopic particles like atoms or electrons. And repeating this thousands of times. You will discover that, bafflingly, there are directions on the clock face where the particles never go, and places where they preferentially go. It is all to do with the quantum wave associated with each particle. In some directions on the clock face, the two waves reinforce each other and in other directions they cancel out. The former directions are those in which ricocheting particles are frequently seen and the latter where they are never seen.

This experiment was in fact done in 1927 by physicists Clinton Davisson and Lester Germer in the United States, and by George Thomson in Scotland. Actually, they bounced electrons off a crystal—but it was basically the same idea. They discovered that there were directions in which electrons were common and directions in which they were never seen. This proved that bullet-like electrons can behave like waves. And it earned Thomson and Davisson the 1937 Nobel Prize for Physics.

So, even in the case of simple particle collisions, interference of quantum waves creates peculiar behavior that is never seen in our large-scale everyday world.

But electrons bouncing off a crystal seem very esoteric. What consequences do quantum waves actually have for the real world? Well, they explain the very existence of the atoms of which you are made.

As discussed in Chapter 25, according to the laws of electromagnetism—the high-point of nineteenth-century physics—an electron orbiting in an atom should continually broadcast electromagnetic waves like a tiny radio transmitter. This should continually sap an electron of energy so that it spirals into the nucleus in about a hundred-millionth of the second. In short, atoms should not exist.

Quantum theory comes to the rescue because it reveals that an electron is a spread-out wave and so cannot be compressed down into the nucleus. But, in physics, there is always more than one way of looking at a situation. An electron in an atom can take many possible paths around the nucleus. For instance, it could travel around it in a circle. Or in a square. Or it could fly off to the nearest star, swing around it, and come back to the nucleus. There are an infinite number of possibilities—each of which has a quantum wave associated with it. In a sense, the electron does all of these things at once in exactly the same way that it can be in many places at once.

But here is the amazing thing. If all the infinity of quantum waves were added together, they would cancel each other out close to the nucleus. In other words, there is no probability whatsoever of an electron getting close to the nucleus and collapsing an atom down to nothing.

The fact that atoms exist—that you exist—is down to the fact that electrons can be in many places at once and do many things at once. Without quantum weirdness there literally would be no everyday world.

30.

LOOPY LIQUID

There is a liquid that never freezes
(and can run uphill!)

> "In some sense, what you might have suspected from the first day
> of high-school chemistry is true: the periodic table is a colossal
> waste of time. Nine out of every ten atoms in the universe are
> hydrogen, the first element and the major constituent of stars.
> The other 10 percent of all atoms are helium."
> —SAM KEAN

ONE OF THE MOST peculiar substances known to man is un-doubtedly helium, the stuff that makes your voice sound squeaky and fills Mickey Mouse balloons. When in a liquid form, it never freezes, and it is the only liquid that can actually run up hill.

Helium is the second most common element in creation. In fact, it accounts for one in every ten atoms in the universe. So, it is a surprise that it was unknown on Earth until a century and a quarter ago.

The reason helium was overlooked was because it is both *chemically inert* (or unreactive) and extremely light. Its inertness means it rarely gets trapped in compounds with other elements; and its lightness means that, as soon as it is released into the air, it floats off into space. And space, it turns out, is where the gas was found in the first place.

Helium is the only element to have been discovered on the sun before it was discovered on Earth. The man who spotted it there was Norman Lockyer, who, among other things, wrote the first book on the St Andrew's rules of golf, founded London's Science Museum and launched the international science journal *Nature*, which he edited for its first fifty years. On October 20, 1868, Lockyer pointed his six-inch telescope at the sun from his garden in the South London suburb of Wimbledon and examined the light with a spectroscope. Crossing the spectrum of a solar prominence—a loop of solar material catapulted from the surface of the sun—was a curious yellow line.

The line was observed the same year from India by the French astronomer Pierre-Jules César Janssen. Both Lockyer and Janssen heated various substances in their laboratories in an attempt to reproduce the spectral feature, but neither succeeded. This led Lockyer, in 1870, to make the bold suggestion that the curious line was the fingerprint of an unknown element. He was ridiculed for proposing the existence of helium and had to wait many years before his critics ate their words.

The man who proved Lockyer right and found helium on the earth was the Scottish chemist William Ramsay, the only person to discover an entire group of the periodic table of elements. In March 1895, while examining the spectrum of the gases given off by a uranium mineral called cleveite, Ramsay spotted a mysterious yellow line. Lacking a good spectroscope, he sent gas samples to both Lockyer and William Crookes, a physicist famous for experimenting with cathode ray tubes and believing in psychic phenomena such as telepathy. Within a week, Crookes had confirmed that the gas was the same as the one Lockyer had observed. Lockyer was beside himself with joy as he squinted through the spectroscope at the "glorious yellow effulgence" he had first seen on the sun a quarter of a century before.

Today, helium is used in arc-welding equipment, gas-cooled nuclear reactors, and lasers. Deep-sea divers avoid the dreaded "bends" by breathing a mixture of helium and oxygen. But helium's greatest claim to fame is as the ultimate refrigerant. Boiling at the lowest temperature of any substance—a mere 4.2 degrees above absolute zero, or -269 °C—liquid helium is used to cool anything from astronomical detectors to "superconducting" magnets.[1]

In fact, in its liquid state, helium can lay claim to being one of the most bizarre substances known to science. When cooled below 2.18 Kelvin, it becomes a *superfluid*, able to flow without friction and even to run up hills.

The key to understanding such a superfluid, or *quantum liquid*, is that all the atoms act in lock-step, or are *correlated*. In a sense they act like a single super-atom. When a normal fluid passes over a surface, individual atoms of the fluid bump into atoms on the surface and lose energy. Such friction, or *viscosity*, slows the fluid. However, individual atoms of a superfluid cannot lose energy to surface atoms because they are in a sense lashed together with all the other atoms of the superfluid. The superfluid therefore flows without friction and effectively has zero viscosity.

The friction with a surface is so low in fact that the tiniest nudge—a change in temperature or pressure—is enough to cause the superfluid to flow. This means that a minuscule force is all that is needed to set the lightweight liquid climbing against the force of gravity—flowing uphill.

Helium is also the only substance that never solidifies. A liquid freezes when its atoms move so sluggishly that they form a rigid lattice. *Quantum uncertainty* ensures that helium atoms—even at absolute zero—are too restless to stay in a solid lattice, at least at normal atmospheric pressure. Hell will freeze over long before liquid helium does.[2]

31.

UNBREAK MY HEART

In the future, time might run backwards

"Time will run back, and fetch the age of gold."
—JOHN MILTON, "
ON THE MORNING OF CHRIST'S NATIVITY"

COULD TIME RUN BACKWARDS? Answering this question requires first understanding why time runs forwards. And this is not at all obvious. The problem is that the fundamental physical laws that orchestrate our universe have no preferred direction in time. For instance, an atom can spit out a photon of light, and it can also suck in a photon of light. Consequently, if you were to watch a movie of an atom doing something, you would not be able to tell whether the movie is running forwards or backwards. Both events would appear equally reasonable.

Contrast this with our everyday experience. Say you see a photograph of a castle intact and a photograph of the same castle in ruins. Or an intact egg and a broken egg. Or a person when they are a child and when they are adult. In every case, the direction of time is unmistakable. It is obvious which is the earlier photograph and which the later one. Who ever saw a castle uncrumble, an egg unbreak, or a person grow young?

So, why is the behavior of an atom time-reversible whereas the behavior of a collection of atoms—such as a castle, an egg, or a person—is not time-reversible? It's all to do with order and disorder. What all these scenarios have in common is they describe a transition from order to disorder. A castle in ruins is clearly more disordered than a castle intact. And, fundamentally, it is the change from disorder to order that we associate with the direction, or *arrow*, of time.

The reason big objects—large collections of atoms—tend to go from a state of order to disorder is all down to probability. Imagine the egg again: first intact, and then the same egg shattered into pieces. There is only one way that it can be intact but there are many ways it can be broken. For instance, it can be broken into two pieces, or three pieces, or four, and so on. Even if it breaks into four pieces, one might be a large fragment and the other three small, or two might be big and two small. You get the idea.

Crucially, if each outcome is equally probable, then the egg is overwhelmingly likely to end up broken—because there are simply so many more ways that it can be broken than there are ways it can remain intact.

This is why there is an arrow time: because, for bodies that consist of large collections of components, there are always more disordered states than ordered ones. So, there is always a strong tendency for disorder to increase. In fact, this is a statement of one of the cornerstones of physics: the second law of thermodynamics. As outlined in Chapter 14, *entropy* is the word physicists like to use for disorder; entropy, they say, can never decrease.

Of course, things can get disordered only if they were ordered in the first place. And, if we trace everything back to the beginning, the conclusion must be that the universe itself started out in an ordered state. This is a big problem for physicists because an ordered state is synonymous with a "special" state, and, to a physicist, invoking something special as an explanation for anything is tantamount to invoking God. However, it does appear that the Big Bang in which the universe was born was highly ordered.

So, the ultimate reason that time flows forwards is that the universe began in an ordered state and so has the possibility of becoming ever more disordered. But what if one day it were to shrink back down in a "big crunch," a sort of mirror image of the Big Bang in which all of creation ended up squeezed into a single, super-dense point? Since the universe's final destination would, like the Big Bang, be an ordered state, in shrinking down the universe would become ever more ordered. And the unavoidable consequence of this is that time would run backwards. Stars would unexplode, living things would grow young, and so on.

But what is the likelihood of the universe ending in a big crunch? Well, we know that currently the universe is expanding in the aftermath of the Big Bang 13.82 billion years ago because our telescopes show all the galaxies flying apart from each other. Astronomers believed that, if the universe had sufficient mass, one day in the future the gravity of that mass would slow the expansion to a standstill and reverse it. But, in 1998, scientists discovered *dark energy*, invisible stuff that fills all of space and whose repulsive gravity is speeding up the expansion of the universe (see Chapter 43 for more on this). On the face of it, it would appear that a future big crunch is now unlikely. However, observations show that the dark energy gained control of the universe relatively recently in cosmic history. And, since we have no idea why it "switched on," there is always a chance that it could one day switch off again, raising the possibility that the universe might after all shrink back down in a big crunch.

But here is an interesting thing. Although, in a shrinking big-crunch universe, time will run backwards, so too will the thought processes by which any surviving creatures perceive the world. In exactly the same way that a bowl of soup that is not "not hot," a backward-running universe perceived backwards will appear to be going forwards. This raises an astonishing possibility. Although we have convinced ourselves we are living in an expanding big-bang universe, the truth may be that we are actually living in a shrinking big-crunch universe!

32.

WHO ORDERED THAT?

Nature has triplicated its basic building blocks

"The Ramans do everything in threes."
—ARTHUR C. CLARKE[1]

THE BEAUTY OF LEGO is that an infinite variety of things can be built out of a finite number of basic building bricks. But say the Lego corporation decided to launch another form of Lego with all the same bricks but each one hundreds of times bigger? And then it announced there was to be yet another type with all the same bricks but each one thousands of times bigger? That would be mad, of course. But, actually, it is exactly what nature has done with its own fundamental building blocks.

Normal matter is made of just four basic building blocks: two *leptons* and two *quarks*. The two leptons are the electron and the electron-neutrino. The electron is familiar because it commonly orbits in atoms, but the neutrino is less well-known, mainly because it is so ridiculously unsociable. Although neutrinos are generated in enormous quantities by the sunlight-generating nuclear reactions in the heart of the sun, they *interact* with normal matter so rarely that they fly through the earth as easily as light through a pane of glass.[2]

The two leptons of normal matter are joined by two quarks: the up-quark and the down-quark. These clump together in threes to make the proton and neutron, the principal components of the cores, or nuclei, of atoms. The proton is composed of two up-quarks and one down-quark, and the neutron of two down-quarks and one up-quark. The existence of quarks inside nuclei was demonstrated in the late 1960s and early 1970s when particle physicists fired high-speed electrons into protons. They ricocheted exactly as if they were striking and bouncing off three point-like grains buried deep inside.

Bizarrely, it is impossible to knock a quark out of a proton or neutron and therefore to create a *free quark*. This is because of the peculiar behavior of the strong nuclear force that glues quarks together. Not only is it super-strong but it actually gets stronger the further apart the two quarks are—exactly like a piece of elastic, which resists more vigorously the more it is stretched. Long before two quarks are free of each other, the energy put into stretching the elastic is transformed into the mass-energy of new particles, as permitted by the law of conservation of energy.[3] Specifically, the laws of particle physics cause a quark-antiquark pair to be conjured into existence. Experimenters are faced with having to separate two more quarks. But, in attempting to do that, they will create two more, and so on.

The discovery that pretty much all the matter we see in the universe is made of just four building blocks—the electron, neutrino, up-quark, and down-quark—is a truly extraordinary achievement of science. But, as pointed out before, there is a twist. For a reason nobody understands, nature has decided to triplicate its basic building blocks. Instead of one quartet of particles, there are three, each containing successively more massive versions of essentially the same particles. So, in addition to generation 1, which consists of the electron, electron-neutrino, up-quark, and down-quark; there is generation 2, which consists of the heavier muon, muon-neutrino, strange-quark, and charm-quark; and generation 3, which consists of the even heavier tau, tau-neutrino, bottom-quark, and top-quark.

Bizarrely, neither of the two heavier families plays any role whatsoever in the everyday world. In fact, it takes a large amount of energy to create them, so they were common only in the super-energetic fireball of the Big Bang in the first split-second of the universe's existence. In fact, when the muon—essentially a heavier version of the electron—was discovered in 1936, the American physicist I. I. Rabi famously said: "Who ordered that?" The same could be said of all the duplicates of nature's four basic building blocks.

So, are the three generations of fundamental building blocks all there are—or are there more? Surprisingly, it is cosmology, not particle physics, that hints at an answer. Between one and ten minutes after the birth of the universe, the Big Bang fireball was hot enough and dense enough for protons (hydrogen nuclei) and neutrons to run into each other and stick together to make nuclei of the second-heaviest element, helium. This primordial helium has largely survived until today, and it can be observed throughout the universe. It accounts for about 10 percent of all atoms.[4] However, if there were many more generations of neutrinos, they would have speeded up the expansion of the Big Bang fireball, causing the universe to make more helium than astronomers observe around us today. According to calculations, a universe with 10 percent helium atoms is possible only if there are at most three or four generations of neutrinos. So, there may be a fourth even heavier generation of fundamental building blocks still to be found. However, most physicists would bet against it.[5]

Why nature has chosen to triplicate its fundamental building blocks remains one of the deepest mysteries in physics. No doubt the massive versions of the common quarks and leptons played some critical role in creating the universe we see around us. The hope is that that role will become clear when, eventually, physicists obtain a unified description of the fundamental building blocks and the forces that glue them together—the fabled "theory of everything."

33.

A WONDERFUL THING IS A PIECE OF STRING

The universe may have at least ten dimensions

"Technically, you need the extra dimensions. At first people
didn't like them too much, but they've got a big benefit, which
is that the ability of string theory to describe all the elementary
particles and their forces along with gravity depends on
using the extra dimensions."
—EDWARD WITTEN

ISAAC NEWTON WAS FIRST to realize that, at a fundamental level,
all there is to the universe is particles of matter and the forces that
bind them together. We now know of four fundamental forces, of
which gravity and the electromagnetic force that glues together the
atoms in your body and powers our electrical world are the most familiar. As discussed in Chapter 26, it was Albert Einstein who in 1915
realized something unexpected about one of these forces. As we saw
in the previous chapter, the force of gravity does not actually exist.

As previously mentioned, according to Newton, the force of gravity between the sun and the earth is like an invisible tether between
the bodies that keeps the earth forever trapped around the sun. Einstein begged to differ. He showed that, actually, a mass like the sun
warps the space-time around it, creating a valley, and the earth rolls
around the upper slopes of this valley.[1]

According to Einstein, gravity is a force we have invented to explain our motion through the landscape of four-dimensional space-

time because, as three-dimensional creatures, we are completely unaware of this landscape. This insight intrigued Theodor Kaluza and Oskar Klein, two physicists working independently in the 1920s. They wondered whether the only other fundamental force known at the time—the electromagnetic force—might also be a manifestation of the curvature of a higher dimensional space. Astonishingly, it did indeed appear to be. Kaluza and Klein were able to explain both gravity and electromagnetism as consequences of the curvature of five-dimensional space-time.

We do not of course see a fifth dimension of space. Kaluza and Klein maintained that this is because the extra dimension is curled up much smaller than an atom. Unfortunately, Kaluza and Klein's scheme was mothballed when experiments that probed the cores, or nuclei, of atoms revealed the existence of two more fundamental forces: the ultra-short-range strong nuclear and weak nuclear forces.

Fast-forward half a century. Various aspects of the strong force indicated that the fundamental building blocks of matter—six quarks and six leptons—which are glued together by the nature's four fundamental forces, might not actually be particles like tiny billiard balls but one-dimensional strings of mass-energy. The idea is that different vibrations of such strings correspond to different fundamental particles: sluggish, low-energy vibrations correspond to light particles, and rapid, high-energy vibrations to massive particles. Picture the vibrations of a violin string. According to string theory, in the final analysis, physics is music.

But, just as the building-block particles correspond to string vibrations, so too do the particles that transmit the forces. Quantum theory, the hugely successful theory of atoms and their constituents, has shown that the fundamental forces arise from the exchange of force-carrying particles—in the case of the electromagnetic force, the photon.[2] And here the idea of extra space dimensions raises its head again—though in a slightly different form to Kaluza and Klein's concept. In order to explain nature's four fundamental forces, a total of ten dimensions of space-time are needed—that is, not one but six extra space dimensions in addition to the three we are familiar with. Because we observe none of them, string theorists, just like Kaluza and Klein before them, claim they are compactified, or curled up much smaller than an atom.

If the existence of six extra space dimensions, none of which we are aware of, is not too much to stomach, then consider the strings themselves. They are ridiculously, mind-bogglingly tiny—a million billion times smaller than a hydrogen atom—making them essentially unobservable. Consequently, we cannot probe them and directly confirm or refute the theory.

You might wonder why there is such enthusiasm for a theory with unobservable consequences that predicts that we live in a ten-dimensional world when we so evidently appear to live in a four-dimensional one. The answer has to do with the two towering achievement of twentieth-century physics—Einstein's theory of gravity, also known as the general theory of relativity, and quantum theory. The former reigns supreme on the large scale—that of the universe—and the latter on the small scale—that of atoms and their constituents. However, once upon a time, in the Big Bang, the large-scale universe was small-scale—smaller than an atom. To understand the origin of the universe, we therefore need to unite the theory of the big with the theory of the small. String theory is so far the only framework that does this. The reason is that one of the vibrating strings—actually a loop of string—appears to be a graviton, the hypothetical quantum carrier of the gravitational force. String theory, a fundamentally quantum theory, therefore contains within in it a theory of gravity.

34

NO TIME LIKE THE PRESENT

The concept of a common past, present, and
future simply does not exist anywhere in our
fundamental description of reality

"Now he has departed from this strange world a little ahead
of me. That means nothing. People like us, who believe in
physics, know that the distinction between past, present,
and future is only a stubbornly persistent illusion."
—ALBERT EINSTEIN[1]

OUR FUNDAMENTAL PICTURE OF reality is Einstein's theory of relativity. It reveals that how fast time passes for you relative to someone else depends on how fast you are moving relative to them and the strength of gravity you are both experiencing. The first effect is a consequence of Einstein's special theory of relativity of 1905 and the second of his general theory of relativity of 1915. Since all of us on Earth experience pretty much the same gravity, the former effect is the significant one.

The fact that people's experience of time depends on how fast they are moving stems from the fact that, in our universe, for some unknown reason, the speed of light plays the role of infinite speed. As Einstein said: "The velocity of light in our theory plays the part, physically, of an infinitely great velocity."[2]

Just as nothing can catch up something moving at infinite speed, nothing can catch up with a light beam. If something is moving at infinite speed, no matter how fast you are moving your speed will be so tiny in comparison that you will still measure that thing moving at infinite speed. Now substitute light for that "something," and you will understand that the speed of light always appears the same to everyone, no matter how fast they are moving.

Consider how weird this is. The velocity of anything is the distance it travels in a given time—for instance, a car may speed down the motorway at one hundred kilometers an hour. But, if everyone is to measure the same speed for a beam of light, something must happen to each person's ruler and clock. Einstein figured out exactly what. In a nutshell, rulers shrink in the direction of motion and moving clocks run slow. So, if someone is moving past you, they will appear to shrink in the direction they are traveling and move in slow motion! You never actually see this effect because it is apparent only at speeds approaching the speed of light, which is about a million times faster than the speed of sound. However, by this huge cosmic conspiracy in which space and time stretch like elastic, the speed of light is kept the same for everyone.

But, if time flows at a different rate depending on how fast people are moving relative to each other, it follows that there is no such thing as a common past, present, and future. So why do we have such a powerful feeling that there is? Part of the answer is that we live out our lives in the universe's ultra-slow lane and so never see anyone pass us at a speed even close to that of light.

But this still does not explain why we experience a "present." Why do we focus on the information most recently gathered by our senses? Why, for instance, do we not have a delayed present and focus on information from ten seconds ago? Or two presents, and focus on information gathered, say, ten seconds and thirty seconds ago?

Physics provides no explanation for our experience of "now." For this reason, some have looked for other explanations. One such physicist is Jim Hartle of University of California at Santa Barbara, a collaborator of British physicist Stephen Hawking. He thinks that, early in the evolution of life, there may have been organisms that experienced time in myriad different ways, not just the way we do today. But say a tree frog had a delayed present and focused on information from ten seconds ago—if a fly alighted on a leaf in front of it, by the time the tree frog lashed out its tongue, the fly would have long gone. Reliant on such out-of-date information, says Hartle, the tree frog would eventually starve to death.

And this would be true of organisms experiencing time in every way except, crucially, the way we do—that is, by focusing on the most recent information pouring into our senses. So, according to Hartle, rather than there being a physical explanation for the way in which we experience time, there may be a biological one. And, since the same constraints must apply to any life, Hartle concludes that extraterrestrial creatures, wherever they arise in the universe, will likely experience time exactly the same as us.

The person who made the deepest, most insightful, most profound observation about time was ex-US-president George W. Bush: "I think we can all agree that the past is over." Genius.

35.

HOW TO BUILD A TIME MACHINE

Time travel is not ruled out by the laws of physics

> "If you go flying back through time and you
> see somebody else flying forward into the future,
> it's probably best to avoid eye contact."
> —JACK HANDY

NOT ONLY IS TIME travel not ruled out by the laws of physics but the laws of physics actually appear to make it straightforward—at least in principle. It is all down to Einstein. In his general theory of relativity, he showed that time flows at different rates in different gravity: slower in strong gravity, faster in weak gravity.[1] So, all you need in order to make a time machine is a region of space where gravity is weak and time flows normally—say, the earth—and one where gravity is strong and time flows more slowly—say, the vicinity of a black hole.[2]

Now, imagine a clock on the earth and a similar one at the black hole and that both start ticking on Monday. By the time it is Friday on Earth, it is still only Wednesday at the black hole. So, if there is a way to go instantaneously from the black hole to the earth, it is possible to go from Friday back to Wednesday. Remarkably, there is such a way. Einstein's theory doesn't just permit the existence of black holes; it also allows a tunnel, or shortcut, through space-time known as a *wormhole*.

The recipe for a time machine is therefore: take the earth and a region near a black hole and connect them with a wormhole. But there is a snag (isn't there always?). Wormholes are unstable. They have the annoying habit of snapping shut in the merest blink of an eye. To stop this, they must be propped open by stuff with repulsive gravity—stuff that blows rather than sucks. Remarkably, such "exotic matter" exists. In fact, it is the major mass component of the universe, accounting for about two thirds of the mass-energy of the cosmos.

In 1998, astronomers found that the expansion of the universe—which had been kick-started by the Big Bang and should, after 13.82 billion years, be running out of steam—was not running out of steam. It was speeding up. To explain this anomalous cosmic acceleration, they were forced to postulate the existence of dark energy, which was mentioned in Chapter 34. We know it is invisible, fills all of space, and has repulsive gravity. And it is the repulsive gravity that is speeding up cosmic expansion.

But, although dark energy has repulsive gravity, it is far too dilute and feeble to prop open a wormhole. We need stuff like it but with enormously stronger repulsive gravity. In fact, to prop open a wormhole wide enough for, say, a person to crawl through would require stuff with an energy equivalent to that pumped out by an appreciable fraction of the stars in our galaxy during their lifetimes.

There are two key differences between such a time machine and, for instance, the one depicted by H. G. Wells in his novel *The Time Machine*. The hero in the film, played by Rod Taylor, sits in a contraption with brass dials and pulls a lever one way to go backwards in time and the other way to go forwards in time. But Einstein's theory tells us that sitting on the spot is not an option. To travel through time, it is necessary to travel through space too. The second difference between an Einsteinian time machine and Wells's fictional one is that it is impossible to travel back to a time before your time machine was built. So, if you want to go on a dinosaur safari, you will have to find a time machine abandoned on Earth by extraterrestrials sixty-six million years ago!

To recap, then, the ingredients for a time machine are a black hole; a wormhole; a type of matter with repulsive gravity that we do not know exists; and the energy emitted by an appreciable fraction of the stars in our Milky Way during their lifetimes. Nobody said making a time machine was easy!

The point is not that it is possible to build a time machine *in practice*. It would be mind-bogglingly hard. In fact, it would appear impossible, except for a super-advanced technological civilization. No, the point is that it is possible to build a time machine *in principle*. It is this fact that keeps physicists awake at night because it opens the door to all sorts of crazy things. For instance, if a time machine existed, someone could use it go back in time and shoot their grandfather before their mother was born. Admittedly, it might not be anything most people would want to do, but it takes only one person. And that possibility is enough to alarm physicists. Because they have to confront the question: how did the person go back in time and bump off their grandfather if they had never been born?

To avoid the Grandfather Paradox, as it is known, Stephen Hawking proposed the Chronology Protection Conjecture. It is really just a fancy way of saying: time travel is impossible. In other words, some as-yet-unknown law of physics must intervene and prevent time travel and its associated paradoxes. Hawking's argument is a simple observational one: "Where are the time travelers from the future?"

There is another possible way out of the Grandfather Paradox but it means accepting something very weird about our universe. According to quantum theory, before an atom registers its presence in an experiment, it can exist in many places at once. Physicists therefore have to explain why, when they actually observe an atom, they find it in only one location. To do this, more than a dozen interpretations of quantum theory have been devised, all of which predict the same outcomes in experiments. The most famous is the Copenhagen interpretation, in which it is the act of observing an atom that forces it to behave and be in one place only (the Copenhagen interpretation is itself open to interpretation since it does not specify whether the "observer" is a piece of equipment or a human being).

But, in the Many Worlds Interpretation, an atom is fundamentally in many places at once but in different, or parallel, realities. Since we can experience only one reality, we see only one version of the atom.

If the Many Worlds Interpretation is correct, then the Grandfather Paradox is neatly sidestepped. It is possible for someone to go back in time and kill their grandfather before their mother was born because it is not actually their grandfather. It is a grandfather in a parallel reality!

PART SIX

EXTRATERRESTIAL THINGS

36.

OCEAN WORLDS

Beneath the ice of Jupiter's moon Europa is
the biggest ocean in the solar system

"All these worlds are yours, except Europa.
Attempt no landing there."
—ARTHUR C. CLARKE[1]

JUPITER'S ICE MOON, EUROPA, is the second most distant of the gas planet's four giant Galilean moons. Lacking mountains, valleys, and even craters, it looks exactly like a snooker cue ball. In fact, the surface of the moon is so smooth that if it had an atmosphere, it would rank as the solar system's biggest ice rink. However, while Europa from afar appears to be featureless and dull, close-up it is a very different world.

In 1979, NASA's *Voyager 2* space probe flew through the Jovian system. The images it relayed back to Earth revealed that Europa's icy surface is covered in a vast and complex network of cracks and ridges. Later, in the 1990s, NASA's *Galileo* probe went into orbit round Jupiter. Its more detailed images of Europa showed a fragmented surface. In fact, it bears a striking resemblance to sea ice in the Arctic that has shattered into jagged pieces and drifted about a bit before gluing back together again. It is a strong hint that beneath Europa's ice there lies an ocean.

The origin of such an ocean is not difficult to understand. Europa orbits Jupiter only slightly further away than Io does. The same tidal stretching forces from the giant planet that have turned Io's rocky innards into molten goo and made it the most volcanic body in the solar system should therefore have liquefied Europa's icy interior.[2] In fact, observations by the *Galileo* probe of the spin of Europa indicate that the moon's interior is rotating at a different speed to its icy crust. This makes perfect sense if the crust is floating on liquid. It appears that under a ten-kilometer-thick layer of ice there is a global ocean perhaps one hundred kilometers deep—the biggest ocean in the solar system.

Is it possible that, at this very moment, there is life down there swimming about in the stygian gloom? At one time, sunlight was considered essential for biology. However, in 1977, a discovery on Earth changed everything. A team led by American oceanographer Robert Ballard was using the mini-submarine *Alvin*. Deep down on the ocean floor, they found mineral-rich, super-hot water erupting from *hydrothermal vents*.[3] Around each vent was an ecosystem of sulphur-eating bacteria and weird tube worms the length of an arm.

Similar hydrothermal vents are likely to exist on the ocean floor of Europa since the moon is being heated by the stretching and squeezing of Jupiter. And it is at least conceivable that, just like on Earth, such vents are powering colonies of living organisms.

But Europa is not the only moon in the solar system with a sub-surface ocean. In 2008, NASA's *Cassini* spacecraft had been orbiting Saturn for two years when it obtained one of the most extraordinary images in the history of planetary exploration. This was of the tiny satellite Enceladus, barely larger than southern England. Spouting from the moon and extending hundreds of kilometers into space, *Cassini* snapped gargantuan fountains of sparkling ice crystals.[4]

There had been earlier hints that Enceladus was not a dead moon. It is the whitest and shiniest body in the solar system, which indicates that the meteoritic dirt and grime it should have accumulated over the ages has been overlain by fresh snow. Then there is the strong suspicion that the ice particles of Saturn's tenuous "E" ring are supplied by Enceladus. Finally, in the moon's southern hemisphere, there are four mint-colored fractures, christened "tiger stripes," which indicate that the surface has undergone movement and so is relatively warm. The tiger-stripe region, incidentally, is the source of Enceladus's ice fountains.

Some of the heat driving the fountains may come from tidal stretching of Enceladus by Dione, a moon that makes one circuit of Saturn for every two circuits made by Enceladus. But Enceladus can eject ice crystals at more than 2,000 kilometers an hour, a speed attained by terrestrial geysers only when there is hot water under pressure. There must, therefore, be another heat source in addition to tidal stretching. All evidence points to that heat source being a warm ocean beneath the ice crust of Enceladus: the smallest ocean in the solar system.

Finding so much activity on such a tiny, cold body is a huge surprise. Everyone thought that the effect of tidal heating would be appreciable only on a big moon. Nobody imagined that it might operate on anything as small as Enceladus. With its subsurface ocean, Saturn's tiny ice moon has turned our ideas about where we might find life on their head. Organic molecules—the ingredients of life—are known to cause the mint-green color of the moon's tiger stripes. That means that, in the warm, wet interior of Enceladus there exist all the ingredients for a second biology. Or maybe—if Europa has pipped Enceladus to the post—a third biology.

Enceladus teaches us that life may exist in the solar system in all kinds of places where we would never have guessed it might be found. If the moon does indeed contain a teeming ecosystem of microorganisms hidden away in darkness since the birth of the solar system, Saturn's "E" ring, fed by Enceladus's ice fountains, may be more than simply water-ice. It may be an orbital graveyard of frozen bacteria.

37

ALIEN GARBAGE

If aliens are out there,
their garbage must be here on Earth

"How would it be if we discovered that aliens only stopped
by Earth to let their kids take a leak?"

—JAY LENO

I N THE FILM *2001: A Space Odyssey*, astronauts excavate an alien
artifact buried beneath the crater Tycho on the moon. When the
sun rises, exposing it to the light of the lunar dawn, it broadcasts an
ear-splitting message to the stars. Three million year ago, its makers
flew through the solar system. They saw the abundant life on the
third planet from the sun and recognized its potential. But they
could not stop. There were so many other planetary systems in the
galaxy to explore. So they buried a baby monitor on the moon,
primed to warn them the day a technological civilization arose on
the third planet and crossed the gulf of space to its giant satellite.

Might this be more than just fiction? Could there truly be alien
artifacts buried beneath the surface of the moon or even the surface
of the earth? According to a radio astronomer in Kharkov in the
Ukraine, the answer depends only on whether there are space-faring
extraterrestrial civilizations in the Milky Way. If there are, says
Alexey Arkhipov, the presence of alien artifacts on our doorstep is

not just possible; it is guaranteed. He goes further. According to his estimates, since the birth of the solar system, thousands of ET artifacts may have fallen to Earth.[1]

The key word here is "fallen." Arkhipov is not claiming that alien artifacts have been left on Earth deliberately as they were left on the moon in the film. Instead, he is asserting that they have fallen to Earth accidentally.

Arkhipov points out that our space activities unavoidably pollute our solar system. Dead satellites, discarded rocket casings, and so on, end up cluttering Earth's orbit. Such space junk is so hazardous that it caused the postponement of launches of NASA's space shuttle for fear of a fatal collision. But, says Arkhipov, such interplanetary garbage does not stay interplanetary garbage forever. Inevitably, some man-made artifacts are ejected from the solar system and sail off towards the stars. Ultra-small particles in the exhausts of space rockets will be blown away by the pressure of sunlight; space probes that explode while far from the sun will eject debris into interstellar space.

But all this works both ways, says Arkhipov. Just as human activities pollute the solar system with garbage, the activities of any space-faring ETs will fill their planetary system with space junk too. And, just as our technological activities lead to the scattering of artifacts into interstellar space, theirs will also. In other words, it is inevitable that some of their junk will come our way. "For Christopher Columbus, the evidence of new lands was strange debris which had floated across the ocean," says Arkhipov. "In the same way, debris which has floated across the ocean of space could provide us with the unmistakeable evidence of new planets and new life."

Arkhipov has even estimated how many alien artifacts of a variety of sizes would have fallen to Earth over its 4.5 billion-year history. Obviously, it involves making some educated guesses. He assumes, for instance, that 1 percent of nearby stars are home to technological civilizations and that, over their history, these civilizations turn 1 percent of the material locked up in their asteroids into the ET equivalent of consumer goods. He believes that is not unreasonable to assume that such civilizations will eventually exploit all the resources in their space neighborhood just as we are currently exploiting all the resources in our neighborhood on Earth.

Arkhipov's conclusion is stunning. "During its 4.5 billion-year history," he says, "the earth could accumulate about 4,000 100-gram artifacts." One hundred grams, by the way, is about the weight of a small jar of Marmite.

Of course, if only 0.01 percent of nearby stars are home to technological civilizations, or they turn only 0.01 percent of the material locked up in asteroids into consumer goods, Arkhipov's estimate of 4,000 artifacts must be scaled down to four.

The evidence of ETs could literally be beneath our feet. Arkhipov says scientists should seriously consider looking for such artifacts in geological strata and among unusual meteorites. Probably, the best place to look is, as Arthur C. Clarke guessed, on the moon. Although it has been struck by meteorites, it has not been weathered or remade by geological forces. Unfortunately, it is currently beyond our capability to undertake such investigations on the moon, so our best bet is our own planet. The most likely place for an alien artifact to end up is in the ocean since ocean covers almost two-thirds of the earth. However, the pressure at the foot of the deepest ocean trenches is so cripplingly high that we can send only robotic emissaries, so finding the odd alien artifact the size of a Marmite jar seems a near-impossible challenge.

A better place to look would be on the land. But, of course, the effect of wind, rain, and ice can over time weather away even the tallest mountains. And even these forces pale into insignificance be-sides the geological ones that, over hundreds of millions of years, open up new oceans and cause continents to dive into oblivion in the magma beneath our feet. The prospects for finding an alien artifact do not look good. In fact, any artifact that fell to Earth more than a billion years ago has probably long ago been dragged down below the earth's surface, there to be crushed and transformed by the heat and pressure of our planet's interior.

Artifacts that fell to Earth more recently, however, might still be near the surface. But, recognizing one would be a challenge, to say the least. An artifact of an advanced extraterrestrial civilization thousands, even millions of years ahead of us might be as unrecognizable to us as a dishwasher is to an ant, or even as an amoeba. As Clarke said: "Any sufficiently advanced technological civilization is indistinguishable from magic."

Our only hope, it seems, would be to find a piece of rock or metal with an unusual chemical composition or even an unusual nuclear composition. Perhaps, at this very moment, somewhere in the world, a puzzling artifact is lying in a dusty museum collection. Maybe nobody has noticed it for a century or more. Or perhaps a new curator is now lifting it from a glass case and looking at it with a furrowed brow.

38.

INTERPLANETARY STOWAWAYS

Want to see a Martian? Look in a mirror.

"The Martians are always coming."
—PHILIP K. DICK

N 1976, LIFE WAS discovered on Mars. Or so some believe. The claim remains deeply controversial. It centers on experiments to detect the signs of biology, which were carried to the Martian surface by NASA's *Viking 1* and *Viking 2* landers. Unfortunately, the experiments yielded ambiguous rather than clear-cut results.

The idea behind the experiments was simple. Scoop up some desiccated Martian soil. Add water, warmth, and nutrients. If the soil contained any dormant, slumbering microorganisms, they should revive and begin metabolizing, generating as a by-product the gas carbon dioxide. Imagine the excitement of the Viking team when its experiments did indeed generate a burst of carbon dioxide. The trouble was, it was hugely bigger than anyone expected, and it died away quickly. It did not look like the signature of biology.

Despite this, the designer of the crucial experiment, ex-Californian sanitation engineer Gilbert Levin, still believes that the *Viking* landers found Martian life. Most other scientists, however, think they discovered weird Martian soil chemistry, perhaps involving highly reactive "peroxides," which quickly oxidized the nutrients to make the carbon dioxide.

On the face of it, the environment on Mars appears too harsh for life. Orbiting 50 percent farther from the sun than the earth, and with an atmospheric blanket—principally of carbon dioxide—only 1 percent as thick as the terrestrial atmosphere, Mars on a hot summer's day barely reaches the freezing point of water. The world also lacks the shield of a planetary magnetic field, which means its surface is exposed to deadly particle radiation from the sun.

But the discovery of hydrothermal vents on Earth made by marine geologist Bob Ballard and his team in 1977 changed our view of what kind of environments can support life. Kilometers down on the sea floor, they gush superheated minerals into the ocean and support a thriving ecosystem, all in total darkness. At the bottom of the food chain are bacteria, which get their energy not from oxygen but from sulphur compounds, and at the top are giant tube worms.

Since Ballard's discovery, biologists have found microorganisms thriving in the most desiccated and desolate wastes of Antarctica. They have found them living several kilometers beneath the earth's surface in solid rock. In fact, one species of bacteria, *Deinococcus radiodurans*, dubbed "Conan the Bacterium," is even happy living in the cores of nuclear reactors. Who then would bet against super-tough bacteria, or *extremophiles*, such as these living on Mars, perhaps in caves beneath the surface where there is permafrost—maybe even running water—and where the rock shields them from solar radiation?

Such organisms may have survived from Mars's distant past. For a striking feature of the Red Planet is that the Mars of today is very different to the Mars of yesteryear. The Valles Marineris, a giant canyon that boasts tributaries each bigger than Earth's Grand Canyon, appears to tell of a time when rivers and floods gouged the Martian surface. There are even hints that, once upon a time, Mars had a shallow ocean.

All of this has implications for us. The reason is that the earth was uninhabitable for half a billion to a billion years after its birth 4.55 billion years ago. Although there is tentative chemical evidence of biology from 3.8 billion years ago, the first evidence of fossil bacteria comes from about 3.5 billion years ago. Crucially, Mars, being smaller than Earth, cooled from its molten birth state more quickly than the earth.[1] Consequently, it would have had rivers and oceans when the earth was still a hell hole. If life got going on Mars, therefore, it would have got going earlier than on our planet.

A dozen or so Martian meteorites have been discovered on Earth. They were blasted into space when asteroids struck the Red Planet. They orbited the sun until they were intercepted by the earth. Experiments have shown that extremophiles inside such rocks could survive violent ejection from a planet, long periods in the cold of space, and entry into a planetary atmosphere. Could it be that, 3.8 billion years ago, a meteorite arrived from Mars and seeded Earth with its first microorganisms? Could it be that we are all Martians?

39.

STARDUST MADE FLESH

You were literally made in heaven

"I believe a leaf of grass is no less than the
journeywork of the stars."[1]
—WALT WHITMAN

THE IRON IN YOUR blood, the calcium in your bones, the oxygen
that fills your lungs each time you take a breath—all were forged
inside stars that lived and died before the earth was born. How we
discovered this remarkable fact—that we are far more intimately
connected to the stars than even astrologers dared imagine—is a
long and torturous story.

First came the discovery that everything is made of atoms. "If in
some cataclysm all of scientific knowledge were destroyed and only
one sentence passed on to succeeding generations, what statement
would convey the most information in the fewest words?" asked
Richard Feynman. He had absolutely no doubt about the answer:
"Everything is made of atoms."

Paradoxically, the fact that nature is ultimately made of tiny indestructible grains that cannot be changed into anything else became
clear only after centuries of failed attempts to change one substance
into another—for instance, lead into gold. But atoms are not only

elemental; they are the alphabet of nature. As previously mentioned, by assembling them in different ways it is possible to make a galaxy or a tree or a mountain gorilla. The complexity of the world is an illusion. Underneath, things are actually simple. Everything is in the permutations of nature's basic building blocks.

There are ninety-two naturally occurring types of atoms or elements—ranging from hydrogen, the lightest, to uranium, the heaviest. Some are common in the universe and others are rare. And the peculiar thing, discovered in the twentieth century, is that there appears to be a connection between how abundant an element is and the properties of the core, or nucleus, of its atoms. For instance, those atoms whose nuclei are most tightly bound tend to be more common than other atoms.

Why should there be a correlation between the abundance of an element and the nuclear properties of its atoms? The only explanation must be that nuclear processes played a role in actually making atoms. In other words, the ninety-two elements were not placed in the universe on Day One by a Creator. Instead, when the universe was young, it contained only the simplest type of atom, hydrogen, and all the heavier atoms have been built up since from this basic building block.

The protons in atomic nuclei repel each other fiercely, so getting them close enough for the nuclear force to grab them like a *Star Trek* "tractor beam" and glue them together means slamming them together at high speed.[2] Since temperature is a measure of microscopic motion, this requires extremely high temperatures.

The question physicists in the twentieth century faced was: where in the universe is the high-temperature furnace where atoms are forged? At first, it looked as if stars were not hot enough. It turned out that this assumption was a mistake. But it shifted the focus to the earliest moments of the universe and the furnace that existed at the beginning of time: the fireball of the Big Bang. However, nature is not that simple. There is not one cosmic furnace responsible for forging its ninety-two elements. The lightest elements, such as helium, were indeed cooked in the first few minutes of the universe's existence. But all heavier elements have been painstakingly built up since the Big Bang inside stars.

Stars like the sun never get hot enough or dense enough to build any elements heavier than helium, the second-lightest atom in nature. But the most massive stars do forge ever heavier elements all the way up to iron.[3] They end up with an internal structure reminiscent of an onion with each successive onion skin made of heavier elements than the one inside it. Of course, if such stars did not eventually become unstable and explode as supernovae at the end of their lives, all of these elements would stay locked away forever, and we would not be here. Fortunately, these stars not only spray interstellar space with the products of their nuclear furnaces but also make even heavier elements in the furious fireball of their explosions. Those elements intermingle with the gas and dust of interstellar clouds, enriching them with heavy elements. Later, when such clouds fragment, those elements are incorporated in newborn stars and their planets. This is how the heavy elements ended up on the earth. As the American astronomer Allan Sandage pointed out: "We are all brothers. We were born in the same supernova."

Would you like to see a piece of a star? Hold up your hand. You are stardust made flesh.

40

THE FRAGILE BLUE DOT

The most amazing picture ever taken of the
earth is just one pixel across

> "I'm sure the universe is full of intelligent life.
> It's just been too intelligent to come here."
> —ARTHUR C. CLARKE

NASA'S *VOYAGER 1* AND *Voyager 2* spacecraft were launched in 1977. A notable feature of the space probes is that each carried a golden disk, of the grooved type played on an old-style record player. Impressed on those disks are sounds and images of life and culture on Earth. The hope is that the disks—in effect, cosmic time capsules—may one day be intercepted by intelligent extraterrestrials or even future humans whose spaceships have reached the Voyager probes. Neither craft is aimed at any star in particular. However, in about 40,000 years' time, *Voyager 1* is expected to pass within 1.6 light-years of the star Gliese 445.

By 1980, *Voyager 1* had flown past Jupiter and Saturn and sent back photographs of the giant planets' clouds and spectacular moons. It had streaked beyond the outermost planet in the solar system and was heading out towards the stars.

Carl Sagan, famous for presenting the *Cosmos* TV series, was first and foremost a planetary scientist. He was also a member of the Voyager team. For many years, he had been urging NASA to turn the

Voyager 1 camera around to look back the way it had come. Finally, on February 14, 1990, he got his way, and the camera was turned towards the inner solar system.

The image captured is iconic in the history of science. It ranks alongside that taken by the Apollo 8 team of the earth rising above the desolate surface of the moon, and the groundbreaking first image of the spiral staircase of DNA. Crossing the blackness of space are several parallel bands of color that often confuse people. They are of no consequence; they are merely artifacts created by light bouncing around inside the Voyager camera. The key thing is the tiny blue dot in the center of the image. It is only one pixel in size.

All seven billion of us live on that dot. All human history has been played out on that dot. In fact, the entire history of life has been played out on that dot. That blue dot is of course the earth. In fact, it is the image of the earth taken from the greatest distance away—6.1 billion kilometers, or forty times further from the sun than the orbit of the earth.[1]

Now and then, I post this image of the pale-blue dot on Twitter with the caption: "How about remembering we are all on this dot together?" Perhaps because it gives people some much-needed perspective on their lives, the tweet invariably gets a bigger reaction than anything else I ever post. But maybe it is not simply perspective. Maybe it reminds us of our cosmic loneliness.

We live in a universe that contains about two trillion galaxies like our Milky Way. And each galaxy contains on average about one hundred billion stars. From looking at our cosmic neighborhood, we now know there are more planets than stars. In fact, there are more planets in the universe than there are sand grains on all the beaches along all the coastlines on Earth. And yet, in all that mind-cringing immensity, there is only one place we know of where there is life…

That tiny, fragile blue dot.

PART SEVEN

COSMIC THINGS

41.

THE DAY WITHOUT A YESTERDAY

The universe has not existed forever—
it was born

"First of all, the Big Bang wasn't very big. Second of all, there was no bang. Third, the Big Bang theory doesn't tell you what banged, when it banged, how it banged. It just said it did bang. So, the Big Bang theory in some sense is a total misnomer."

—MICHIO KAKU

PERHAPS THE GREATEST DISCOVERY in the history of science is that the universe has not existed forever. It was born. There was a day without a yesterday. About 13.82 billion years ago, all matter, energy, space—and even time—burst into being in a searing hot fireball called the Big Bang. The fireball expanded and cooled, and out of the debris there congealed the galaxies—great islands of stars of which our Milky Way is one among an estimated two trillion.

The idea of the Big Bang was not popular with scientists. In fact, they had to be dragged, kicking and screaming, to it. The reason is that it forced them to confront that most awkward of questions: what happened before the Big Bang? But, no matter how uncomfortable it makes them, scientists have no choice but to go where nature's evidence points. And where it points is overwhelmingly to the universe having been born (and not that long ago, either—the universe is only three times older than the earth).

The first piece of evidence for the Big Bang was found by the

American astronomer Edwin Hubble. In 1929, using the giant 100-inch Hooker telescope in Mount Wilson in Southern California, Hubble discovered that the universe is expanding, its constituent galaxies flying apart like pieces of cosmic shrapnel. The obvious implication is that the universe was smaller in the past. In fact, if we imagine the expansion running backwards like a movie in reverse, we come to a time—13.82 billion years ago—when everything in creation were squeezed into the tiniest of tiny volumes. This was the moment of the universe's birth: the Big Bang.

There is just one possible way to avoid this conclusion. In 1948, British astronomers Fred Hoyle, Hermann Bondi, and Tommy Gold proposed that as the galaxies recede from each other, new material fountains into existence in the gaps between them, eventually congealing into new galaxies. This is a way in which the universe can expand, as Hubble discovered, while having no beginning. This continuous creation of matter might seem ridiculous, but, really, it is no more ridiculous than the creation of all matter in a one-off Big Bang.

The most important aspect of Hoyle, Bondi, and Gold's steady state theory was that it was testable. It predicted that the universe must look the same at all times. However, in the 1960s, astronomers discovered *quasars*, the super-bright cores of newborn galaxies.[1] Their light had taken many billions of years to travel across space to the earth and so was showing them they were in the early universe. Since there are no quasars in existence today, it was obvious that the universe has changed, or evolved, flatly contradicting the steady state theory.

But the killer blow to the steady state theory came in 1965. At Holmdel in New Jersey, two astronomers were using a giant horn-shaped radio antenna.[2] It had been built at Bell Labs, part of the AT&T phone company, to transmit and receive microwave signals from the first experimental communication satellites. Arno Penzias and Robert Wilson aimed to use it to detect the faint radio-glow of ultra-cold hydrogen gas, which they believed was surrounding our Milky Way. However, their project was frustrated by a persistent microwave hiss of static they picked up wherever they pointed their horn in the sky. Inadvertently, they had stumbled on the heat afterglow of the Big Bang fireball. Greatly cooled by the expansion of the universe in the past 13.82 billion years, it appears today not as high-energy visible light but as low-energy microwaves.

The *cosmic background radiation* not only earned Penzias and Wilson the 1978 Nobel Prize for Physics but also confirmed the Big Bang. Despite this, Hoyle—who was, ironically, responsible for coining the term "Big Bang" in a BBC radio broadcast in 1949—never accepted that Penzias and Wilson's microwaves came from the beginning of time; and until his death, he concocted ever more elaborate ways to incorporate the cosmic background radiation into a modified steady state theory.

But, actually, the fact that the universe must be expanding and therefore must have had a beginning was actually deduced independently by two physicists: the Russian Alexander Friedmann in 1922 and the Belgian George Lemaître in 1927.[3] The Big Bang universe—also referred to as a Friedmann-Lemaître universe—particularly appealed to Lemaître because, in addition to being a scientist, he was a Catholic priest. It seemed to him that a universe that burst into existence in a bright fireball was perfectly compatible with the God of Genesis creating the universe by saying: "Let there be light: and there was light."

Friedmann and Lemaître actually deduced the expanding universe from Einstein's theory of gravity—the general theory of relativity—which Einstein presented to the world at the height of the

First World War in 1915. But, although Einstein the following year applied his theory to the biggest gravitating mass he could imagine—the entire universe—he missed the message in his own equations. This is a common mistake among scientists. It is incredibly difficult for them to believe that the universe really dances to the tune of the arcane formulae they scrawl across blackboards. As Nobel Prize-winning physicist Steven Weinberg has observed: "The mistake of physicists is not in taking their theories too seriously but in not taking them seriously enough."

The birth of the universe in a fireball leaves science with a huge challenge. "Every time we get a story that says there was a Big Bang, then people want to know what was before that," says John Mather, who won the Nobel Prize for his space-based observations of the cosmic background radiation. "And if we find out, what was before that?"[3]

"We can trace things back to the earlier stages of the Big Bang, but we still don't know what banged and why it banged," says astronomer Royal Martin Rees. "That's a challenge for twenty-first-century science."

42.

GHOST COSMOS

The universe we see in our telescopes is
not actually out there

"Space is big. You just won't believe how vastly, hugely, mind-
bogglingly big it is. I mean, you may think it's a long way down
the road to the chemist's, but that's just peanuts to space."

—DOUGLAS ADAMS[1]

IMAGINE YOU LIVE IN Central London and you look out of your
window and one hundred meters away horse-drawn carts are clog-
ging the streets. Three hundred and fifty meters away, the Great Fire
of London is turning the sky ruby-red. Two kilometers away, the first
Roman ships are docking on the marshy banks of the River Thames.
Ridiculous? Well, this is exactly the kind of thing astronomers experience when they look out across the universe with their telescopes.

What I have described is simply what you would see from a window in Central London if light was slowed to a speed of one hundred
meters per century and so brought news of events at a snail's pace.
The speed of light is of course enormous—about 300,000 kilometers
per second. However, the expanse of space light must span to reach
the earth is truly vast. Consequently, it appears to crawl across it at
the pace of a cosmic snail.

The further across the universe we see, the further back in time
we probe. We see the moon as it was 1.25 seconds ago, the sun as it

was 8.5 minutes ago, and the nearest star system, Alpha Centauri, as it was 4.25 years ago. It is literally impossible to know what the universe looks like at this moment. In fact, the concept of "now" is meaningless in our cosmos.

We can be reasonably sure that the moon, the sun, and the nearest star system are still there—and probably the nearest galaxy, Andromeda, which we see as it was 2.5 million years ago. But this may be not true of galaxies we see as they were many billions of years ago. They may have long ago died, their stars having winked out, their hearts having been cannibalized by other galaxies. Take, for example, quasars, which derive their phenomenal light output from matter heated to incandescence as it swirls down on to a "supermassive" black hole.[2] Quasars long ago exhausted their food supply of gas and ripped-apart stars. So no quasars exist in today's universe. When they pop up in our telescopes, they are like a persistent afterimage of a super-bright firework that long ago faded and died.

The snail-like progress of light across our enormous universe essentially transforms telescopes into time machines. And, here, what nature takes away with one hand it gives back generously with the other. For although we cannot know what the universe is like "now," by looking further and further out into space we can see what the universe was like at successively earlier epochs. It is an ability historians and archaeologists would kill to have, and it permits astronomers to see the evolution of the universe all the way from the Big Bang to the present day.

But there is a further twist. Not only is much of the universe that we see in our telescopes no longer there but, when it was there, it was not the way it appears to us. This is because the light of distant galaxies, on its long journey across space to the earth, passes by more nearby galaxies. And the gravity of those foreground galaxies bends and distorts the light of the more distant galaxies. It is a phenomenon known as *gravitational lensing*, and it means that much of what we see is distorted, like looking through the frosted glass of a bathroom window. Not only do we live in a ghost universe but even the ghosts we observe are not what they seem!

43

HEART OF DARKNESS

97.5 percent of the universe is invisible

"Dark matter is everywhere. In this room. Everywhere."
—FABIOLA GIANOTTI

THE UNIVERSE IS 97.5 percent invisible. Any way you look at it, this has to be one of the most shocking discoveries in the history of science. However, it has yet to trickle into the consciousness of most scientists. Many have not woken up to the realization that all they have been studying—everything that has been the focus of science for more than 350 years—is no more than a minor contaminant of the universe, like the dusting of snow on a mountain.

About 4.9 percent of the universe is made of atoms—the stuff of which you and I and the stars and galaxies are made. And of that, we have seen only half directly with our telescopes. Astronomers suspect that the other half is gas drifting between the galaxies, either too cold or too hot to glow with light. In fact, there are recent claims that some but not all of this missing stuff has actually been found in the guise of hot filaments of gas, forming a kind of web of tenuous matter between the galaxies.[1]

But, in addition to the 4.9 percent ordinary matter, 26.8 percent—

almost six times as much—is in the form of dark matter. This gives out no light or at least too little light to be detected by our most sensitive astronomical instruments. We know of its existence only because its gravity tugs at the visible stars and galaxies, causing them to move in a manner different from the prediction of Newton's law of gravity.

As for the identity of the dark matter, your guess is as good as mine. Speculations range from as-yet-undiscovered subatomic particles to fridge-sized black holes surviving from the earliest moments of the Big Bang to relics from the future in which time runs backwards (seriously!).[2] If the dark matter is made of the former, it may at this moment literally be in the air all around you. There was a hope that a candidate subatomic particle might turn up at the Large Hadron Collider, the giant particle accelerator near Geneva in Switzerland. But so far, no joy. In idle moments, I daydream about whether there might not be dark stars, dark planets, and dark life, and that the real reason a fifty-year search for extraterrestrial intelligence has drawn a blank is that the dark stuff is where all the action is, with the chaos of galactic commerce going on all around us. Invisibly.

But, in addition to the 4.9 percent ordinary matter and the 26.8 percent dark matter, a whopping 68.3 percent of the mass of the universe is dark energy (remember: all energy has an equivalent mass—it weighs something—according to Einstein's famous $E = mc^2$ formula). The dark energy is invisible, fills all of space, and has repulsive gravity. The repulsive gravity is speeding up the expansion of the universe, which is how the dark energy was discovered in 1998. That's not long ago at all. Just imagine. Until only a couple of decades ago, science had overlooked this major mass component of the universe.

Now, if physicists are baffled by dark matter, they are utterly at sea when it comes to dark energy. Our very best theory of physics is quantum theory. It is fantastically successful. It has given us lasers and computers and nuclear reactors. It explains why the sun shines and why the ground beneath your feet is solid. But when quantum theory is used to predict the energy of the vacuum—that is, the energy of dark energy—physicists get a number that is one followed by 120 zeroes bigger than what we observe. This is the biggest discrepancy between a prediction and an observation in the history of science. I don't think it is controversial to say there is probably something a bit wrong with our understanding of reality.

"Perhaps the most embarrassing aspect of our modern cosmology is the dominance of invisible components," says American astronomer Stacy McGaugh. "Dark matter and dark energy comprise about 95 percent of the mass-energy content of the universe, yet we have only ideas about what they are."[3]

It is a sobering thought that we have constructed a picture of the universe—the great edifice of modern cosmology—on the basis of the mere 2.5 percent of the cosmos we can see directly with our telescopes. Imagine if, in the nineteenth century, Charles Darwin knew about frogs but nothing of trees, dogs, grasshoppers, or sharks. How successful would he have been in coming up with a viable theory of biology like the theory of evolution by natural selection? But cosmologists find themselves in just this position. It seems abundantly clear that some huge idea is missing. Hopefully, when it is found, it will help weld dark matter and dark energy—messy bolt-ons to the basic Big Bang theory—into an elegant, seamless theory. Expect big surprises on the road ahead—surprises that may completely change our view of the universe.

44.

AFTERGLOW OF CREATION

99.9 percent of the photons in the universe do
not come from stars or galaxies but are the
leftover heat of the Big Bang

"Think of such civilizations, far back in time against the
fading afterglow of creation, masters of a universe so young
that life as yet had come only to a handful of worlds. Theirs
would have been a loneliness of gods looking out across
infinity and finding none to share their thoughts."
—ARTHUR C. CLARKE[1]

THE FIREBALL OF THE Big Bang was like the fireball of a nuclear explosion. But the heat of a nuclear fireball dissipates into the surroundings in an hour, a day, a week. By contrast, the heat of the Big Bang had nowhere to go. It was bottled up in the universe, which, by definition, is all there is. Consequently, the heat of the Big Bang is still around us today. It has been greatly cooled by the expansion of the universe in the past 13.82 billion years, so it no longer appears as light visible to the eye but as a type of invisible light known as *micro-waves*.[2]

Microwaves are used by your mobile phone to communicate with other mobile phones, to heat food in a microwave oven, and to transmit TV pictures. In fact, if you could get hold of an old analogue TV and tune it between the stations, about 1 percent of the static, or snow, on the screen would be the afterglow of the Big Bang (this would be equally true of the static of a long-wave radio tuned between the stations). Before being intercepted by your TV aerial, those microwaves had traveled for 13.82 billion years across empty space, and the last thing they touched was the fireball at the beginning of time.

Remarkably, 99.9 percent of the particles of light, or photons, in the universe are tied up in the heat afterglow of the Big Bang, and only 0.1 percent of them are photons from stars and galaxies. This cosmic background radiation is the single most striking feature of the universe. If you had eyes that could see microwaves instead of visible light, the whole universe—all of empty space—would be glowing white. It would be like being inside a giant light bulb. And yet, as mentioned earlier, the cosmic background radiation was not discovered until 1965, and then entirely by accident.

The problem is that everything around us glows with microwaves, making the stuff from the Big Bang hard to spot. This difficulty was faced by two scientists at Holmdel, New Jersey, in 1964. Arno Penzias and Robert Wilson had been lured to Bell Labs by the prospect of using a giant microwave horn for doing astronomy. Penzias and Wilson wanted to use the horn for detecting super-cold hydrogen gas, which they suspected might surround our Milky Way galaxy. Because they expected the microwave signal from this gas to be extremely weak, they needed to first measure the microwave signals coming from all the other sources—nearby buildings, trees, the sky, even the metal of the horn itself—so they could subtract them and be left, hopefully, with only the signal they were looking for.

When Penzias and Wilson did their subtraction, however, they found they were left with a persistent hiss of static. It was exactly what would be radiated by a body at three degrees above absolute zero, or a chilly -270°C.[3] At first, the two astronomers thought they might be picking up microwaves from New York City, which was just over the horizon from Holmdel. But, when they swung the open mouth of their horn around the sky and away from the direction of New York City, the hiss remained unchanged. Next, they thought they might be picking up a source of microwaves in the solar system such as Jupiter, which is known to emit radio waves. But, as the months passed, and the earth traveled around the sun in its orbit, the hiss did not vary. Penzias and Wilson even thought the hiss might come from high-speed electrons injected way up into the atmosphere by a recent nuclear-bomb test. But the hiss did not die away with time, as would be expected.

Finally, the astronomers' eyes alighted on two pigeons, which had made a nest inside the horn at its narrow end. The electronics for detecting microwaves was housed in a small cabin bolted on to the end of the horn. Since the electronics had to be refrigerated, and the refrigerator shed its waste heat, the location the pigeons had chosen was cozy and warm—perfect for the icy New Jersey winter. Penzias and Wilson noticed that the pigeons had coated the interior of the microwave horn with a white *dielectric material*, more commonly known as pigeon droppings. Could this be glowing with microwaves and so causing the persistent hiss of static?

The astronomers trapped the pigeons, posted them in the company mail to another site, and got into the horn with Wellington boots and stiff brooms and scraped away the pigeon droppings.[4] But, to their dismay, when they had finished, they found that the persistent hiss was still there.

By now it was the spring of 1965, and they had done no astronomy whatsoever. It was then that Penzias happened to telephone a fellow scientist. It was about another matter entirely, but he could not help complaining about the trouble he and Wilson were having at Holmdel. Immediately, the other scientist sat up. He had been to a lecture by a theorist called Jim Peebles who talked about an experiment being built at Princeton University—only thirty miles from Holmdel—to pick up the heat afterglow of the Big Bang. When Penzias finished talking, he immediately got on the phone to Peebles's boss, Bob Dicke, at Princeton. At the time, Dicke was having a bag lunch in his office with his research team. When Dicke put down the phone after talking to Penzias, he turned to his colleagues and said: "Well, boys, we've been scooped!"

The radiation discovered by Penzias and Wilson is now known to correspond to a temperature of 2.726 degrees above absolute zero. "The radiation left over from the Big Bang is the same as that in your microwave oven but very much less powerful," said Stephen Hawking. "It would heat your pizza only to minus 271.3 °C—not much good for defrosting the pizza, let alone cooking it!"

For the discovery of the cosmic background radiation, which confirmed the universe had been born in a Big Bang, Penzias and Wilson shared the 1978 Nobel Prize for Physics. And the pigeons? They returned to the Holmdel horn—they were homing pigeons, after all—and, sadly, had to be shot. Their droppings are invariably mentioned in astronomy books. It is almost certainly true that never in the history of physics has something so profound been mistaken for something so mundane.

45.

MASTERS OF THE UNIVERSE

There is a supermassive black hole lurking
like a black widow spider in the heart of every
galaxy—and nobody knows why

"The black holes of nature are the most perfect macroscopic
objects there are in the universe: the only elements in their
construction are our concepts of space and time."
—SUBRAHMANYAN CHANDRASEKHAR

TWENTY-SEVEN THOUSAND LIGHT YEARS away in the dark heart of our Milky Way lies a supermassive black hole 4.3 million times the mass of the sun.[1] Impressive as it is, however, Sagittarius A* is an insignificant tiddler compared with its fifty billion-solar-mass cousins lurking in the cores of some galaxies. The big question is: what are they doing there?

A black hole is a region of space-time where gravity is so strong that nothing, not even light, can escape—hence its blackness. Black holes are a prediction of the general theory of relativity, Einstein's theory of gravity. They are surrounded by an "event horizon," an imaginary membrane that marks the point of no return for in-falling matter and light. Inside the horizon, the distortion of time is so great that time and space actually swap places. This is why the *singularity*—the point at the center of the hole at which in-falling matter is crushed out of the existence—is unavoidable. Since it exists not across space but across time, it can no more be avoided than you can avoid tomorrow.

Once upon a time, black holes were considered more science fiction than science. Even Einstein, whose theory predicted their existence, never believed in them. However, the first stellar-mass black hole, Cygnus X-1, was discovered by NASA's *Uhuru* X-ray satellite in 1971. But, actually, evidence for a far more impressive species of black holes had been found eight years earlier.

In 1963, Dutch-American astronomer Maarten Schmidt discovered quasars, which turned out to be the super-bright cores of newborn galaxies. So distant that their light has taken most of the age of the universe to reach us, quasars are beacons at the beginning of time. A typical quasar pumps out the energy of one hundred normal galaxies like the Milky Way but from a region smaller even than our solar system. Nuclear energy—the power source of the stars—is woefully inadequate. The only conceivable energy source is matter heated to millions of degrees as it swirls on down to a black hole. But not a mere stellar-mass black hole; one with a mass of billions of suns.

For a long time after Schmidt's discovery, astronomers thought supermassive black holes were a cosmic anomaly powering only *active galaxies*—the badly behaved 1 percent of galaxies of which quasars are the most extreme examples. But NASA's Hubble Space Telescope, launched into Earth's orbit in 1990, showed that this is wrong. With its super-sharp vision, Hubble could see and measure the speed of stars swirling round in the hearts of hundreds of galaxies. It was able to show that supermassive black holes are doing the swirling not just in 1 percent of galaxies but in pretty much all of them. It is just that in most galaxies the black holes are quiescent, having exhausted their fodder of interstellar gas and ripped-apart stars. Like our own Sagittarius A*, they are slumbering.

How did a supermassive black hole get to be in the heart of pretty much every galaxy? Did they form after their parent galaxies? Or did they "seed" the formation of galaxies? These are some of the biggest unanswered questions in cosmology.

Stellar-mass holes are believed to form from the catastrophic shrinkage of dying stars, but the formation of supermassive black holes is a total mystery. Perhaps they form when stellar-mass black holes collide and coalesce in the crowded heart of a galaxy. Or maybe they form directly from the shrinkage of a giant gas cloud. The trouble is that astronomers observe supermassive black holes that have already reached billions of solar masses by the time the universe is about 4 percent of its current age—a mere five hundred million years after the Big Bang. It is hard to imagine how they could have grown so big so fast.

But, although a supermassive black hole appears impressive on a human scale, it is extremely small compared with its parent galaxy, and it has a very small mass compared with the mass locked up in a galaxy's stars. The surprise, however, is that, everywhere, supermassive black holes have left their indelible imprint on their parent galaxies. For instance, the mass of the stars in the core of the galaxy is commonly about 1,000 times the mass of the black hole. Clearly, there is an intimate connection between a supermassive black hole and its galaxy. However, it is rather like something as small as a bacterium orchestrating the building of something as big as New York City!

The means by which such tiny supermassive black holes project their power over vast reaches of space are jets. Propelled by a twisted magnetic field in the gas swirling down to oblivion, the jets—channels of super-fast matter—stab outwards from the poles of the spinning black hole. They punch their way through the galaxy's stars and out into intergalactic space where they puff up titanic balloons of hot gas—some of the largest structures in the known universe.

The balloons of gas were, in fact, the first cryptic hint that science got of the existence of supermassive black holes. In the 1950s, radio astronomers, using equipment adapted from wartime radar, discovered that the radio emission observed from some galaxies came not from the central knot of stars, as expected, but, mysteriously, from giant, radio-emitting lobes on either side of the galaxy.

In the early 1980s, the thread-thin jets that are feeding the lobes were imaged for the first time by the Very Large Array of radio dishes in New Mexico. They mock our puny attempts at accelerating matter. Whereas the multi-billion-euro Large Hadron Collider can whip a nanogram or so of matter to within a whisker of the speed of light, nature's cosmic jets can accelerate to similar speeds many times the mass of the sun each year.

The jets help to control the structure of their parent galaxies because, in the inner regions, where they are still fast and powerful, they drive out all the gaseous raw material of stars, snuffing out star formation. However, in the outer regions, where the jets are slower, they slam into gas clouds, causing them to collapse under gravity to give birth to new stars.

But supermassive black holes do not just sculpt galaxies by starting and stopping star formation. According to astrophysicist Caleb Scharf of Columbia University, they may determine the very character of the stars that form. Galaxies with the biggest supermassive black holes—so-called *giant ellipticals*—have a much greater proportion of cool, red, long-lived stars, and there is evidence, says Scharf, that the black hole may be responsible.[2] Such stars spawn planets with few of the heavy elements, such as carbon and magnesium and iron, necessary for life. There is evidence that the kind of chemistry needed for biology might not happen on their surfaces. "Life may have arisen on Earth only because of the Milky Way has a relatively small black hole in its core," he says. "If this had not been the case, the sun and Earth would never have been born."

When we look out across the universe, we see a whole zoo of different galaxies. If Scharf is right, the ones with small supermassive black holes may be filled with planets teeming with life. But others—the ones with the biggest supermassive black holes—may be dead, filled with countless sterile planets.

Black holes have come in from the cold. Though once considered an anomaly, they now appear to play a crucial role in the cosmos. Were it not for the fact that our Milky Way has a modest-sized black hole in its heart, you would not be reading these words.

46.

FLIPPING GRAVITY

Everyone thinks that gravity sucks but in most
of the universe it blows

"I defy gravity."
—MARILYN MONROE

GRAVITY IS A UNIVERSAL force that acts between every chunk of matter and every other chunk of matter. There is a force of gravity, for instance, between you and the coins in your pocket, and between you and a person passing you on the street (though in both circumstances it is far too weak to notice). There is a force of gravity between the earth and the moon, and between the sun and the earth. And, in every case, the force is attractive.

But it does not have to be.

The source of gravity in Isaac Newton's theory is mass. However, in the theory that supplanted Newton's—Einstein's theory of gravity, or the general theory of relativity—it is actually energy. Mass-energy turns out to be merely nature's most compact form of energy.[1] But there are other types of energy—electrical energy, light energy, chemical energy, energy of motion, sound energy, and so on—and all of them have gravity. It seems bizarre that the vibrations of the air that carry your voice have gravity. But, according to Einstein, they really do.

There is twist to this. If you look closely at Einstein's theory of gravity, which he presented in Berlin in November 1915 at the height of the First World War, you will see that things are even more complicated. Yes, the source of gravity is energy—or more specifically, *energy density*, or how concentrated that energy is—but the source is not energy alone: it is energy + pressure.

Now, for a gas in a container, the pressure is simply the averaged-out "push" on the walls of its countless atoms or molecules. In the case of a balloon, for example, the push is provided by billions upon billions of air molecules. It is their incessant drumming on the rubber membrane of the balloon, like raindrops on a tin roof, that keeps it inflated. But in all normal matter the pressure is utterly dwarfed by its energy density. Think of the energy unleashed in the explosion of an H-bomb. That comes from only about a kilogram of matter. Consequently, in all everyday circumstances, pressure can be completely ignored as a source of gravity.

But say that out in the universe there exists a novel type of material that does not exist here on Earth whose pressure is not insignificant compared with its energy density. And say that that pressure is bigger in magnitude than its energy density. And negative. Negative pressure is not anything mysterious. Whereas stuff with normal positive pressure pushes outwards and tries to expand, stuff with negative pressure would suck inwards and try to contract (think of a stretched elastic band trying desperately to shrink). But—and this is the point—if this stuff actually has a pressure that is both negative and bigger in magnitude than its energy density, then the (energy + pressure) term that generates gravity in Einstein's theory becomes a negative number rather than a positive one.[2] In other words, the stuff has repulsive gravity! Rather than sucking, it blows.

Surely the existence of such a ridiculous material is pure science fiction? In fact, it's not.

The universe is expanding, its galaxies flying apart from each other like pieces of cosmic shrapnel in the aftermath of the Big Bang. The only force believed to be operating on the large-scale universe was gravity. Like an invisible web of elastic between the galaxies, it should drag on them and so break cosmic expansion. However, contrary to all expectations, in 1998 astronomers discovered that the expansion of the universe is actually speeding up.

To explain the baffling observation, physicists postulated the existence of dark energy. It accounts for 68.3 percent of the mass-energy of the universe and has repulsive gravity. So, although children are still taught at school that gravity sucks, we now know that isn't the case. In most of the universe, gravity actually blows.

47.

THE VOICE OF SPACE

The black-hole merger detected by its
gravitational waves on September 14, 2015,
pumped out fifty times more power than all the
stars in the universe combined

"If you ask me whether there are gravitational waves or
not, I must answer that I do not know. But it is a highly
interesting problem."
—ALBERT EINSTEIN, 1936

"Ladies and gentlemen, we did it. We have detected
gravitational waves."
—DAVID REITZE, FEBRUARY 11, 2016

NEAR THE TOWN OF Livingston, Louisiana is a four-kilometer-long ruler made of laser light. Three thousand kilometers away in Hanford, Washington State is an identical four-kilometer-long ruler made of laser light. At 5:51 a.m. Eastern Daylight Time on September 14, 2015, a shudder went through the Livingston ruler; 6.9 milliseconds later—less than a hundredth of a second afterwards—an identical shudder went through the Hanford ruler. This was the unmistakable signature of a passing gravitational wave—a ripple in the fabric of space-time itself, predicted to exist by Einstein almost exactly one hundred years ago.

The source of the gravitational waves was an extraordinary event.

In a galaxy far, far away, at a time when the most complex organism on Earth was a bacterium, two monster black holes were locked in a death-spiral. They whirled about each other one last time. They kissed and coalesced. And, in that instant, three times the mass of the sun vanished.[1] It reappeared a split-second later as a tsunami of tortured space-time, propagating outwards at the speed of light.

The power in these gravitational waves exceeded the power output of all the stars in the universe put together by a factor of fifty. Or, to put it another way, had the black hole merger produced visible light rather than gravitational waves, it would have shone fifty times brighter than the entire universe. This is the single most powerful event ever witnessed by human beings.[2]

Gravitational waves are produced whenever mass is accelerated. Wave your hand in the air. You just generated gravitational waves. They are spreading outwards like ripples on a lake. Already, they have left the earth; in fact, they have passed the moon and are well on their way to Mars. In four years' time they will ripple through the nearest star system to the sun. We know that one of the three stars of Alpha Centauri system is orbited by a planet. If that planet happens to be home to a technological civilization that has built a gravitational wave detector, in four years' time it will pick up the ripples in space-time that you made with your hand just a moment ago.

The only problem is that they will be very weak. Imagine a drum. It is easy to vibrate it because a drum skin is flexible. But space-time is a billion billion billion times stiffer than steel. Imagine trying to vibrate a drum skin that is a billion billion billion times stiffer than steel! This is why only the most violent cosmic events, such as the merger of black holes, create significant vibrations of space-time.

But those vibrations, like ripples spreading on a lake, die away rapidly. When they arrived on Earth on September 14, 2015, they had been traveling for 1.3 billion years across space and were fantastically tiny. As they passed the four kilometer rulers at Hanford and Livingston, they alternately stretched and squeezed them—but by only one-hundred millionth of the diameter of an atom! The fact that the twin rulers of the Laser Interferometer Gravitational-Wave Observatory (LIGO) could detect such a small effect is extraordinary.

LIGO is a technological tour de force. At each site there are actually two tubes 1.2 meters in diameter, which form an L-shape down which a megawatt of laser light travels in a vacuum better than interplanetary space. At each end, the light bounces off forty-two-kilogram mirrors suspended by glass fibers just twice the thickness of a human hair and so perfectly smooth that they reflect 99.999 percent of all incident light. It is the microscopic movement of these suspended mirrors that signals a passing gravitational wave. So sensitive is the machine that it was knocked off kilter by an earthquake in China.

To detect gravitational waves, the LIGO physicists had to do something extraordinary: spot a change in length of their 4-kilometer ruler by just 1 part in 1,000,000,000,000,000,000,000. No wonder the 2017 Nobel Prize was awarded to three of the physicists who had pioneered the experiment: Rainer "Rai" Weiss, Kip Thorne, and Barry Barish.[3]

The significance of directly detecting gravitational waves cannot be overstated. Imagine you have been deaf since birth then suddenly, overnight, are able to hear. This is how it is for physicists and astronomers. For all of history they have been able to *see* the universe. Now, at last, they can *hear* it. Gravitational waves are the voice of space. It is not too much of an exaggeration to say that their detection is the most important development in astronomy since the invention of the astronomical telescope by Galileo in 1609.

On September 14, 2015, at the very edge of audibility, we heard a faint sound like the rumble of distant thunder. But we have yet to hear the gravitational-wave equivalent of a baby crying or music playing or a bird singing. Over the next few years, as LIGO increases its sensitivity and other detectors come online in Europe, Japan, and eventually India, our ability to detect gravitational waves will get better. And who knows what we will hear as we tune into the cosmic symphony?

48.

POCKET UNIVERSE

You can fit the information for 64 million
universes on a single 64GB flash drive

"And I say to any man or woman,
Let your soul stand cool and
composed before a million universes."
—WALT WHITMAN, "SONG OF MYSELF"

THE UNIVERSE IS EXPANDING, its constituent galaxies flying apart like pieces of cosmic shrapnel in the aftermath of the Big Bang. This means that, if the expansion is imagined running backwards, the universe gets ever smaller. As seen in previous chapters, the universe is also quantum—which means it is not only fundamentally unpredictable but also grainy. Everything comes in *quanta*—indivisible grains that cannot be cut any smaller: matter, energy, even space. So, if you could see space on the smallest scale with some kind of super-microscope, it would look a bit like a wavy chessboard with squares that could not be made any smaller.

Now, if we imagine the space shrinking as we run the expansion of the universe backwards, the chessboard gets smaller, but the chess squares cannot shrink. So there are fewer and fewer of them. In fact, near the beginning of the universe, at a time known as the *inflationary epoch*, there were only about 1,000 chess squares. That's only 1,000 places in which to either put energy or not put energy. If you are into computers, you will understand that this means the universe during the inflationary epoch was describable by only 1,000 binary digits (0s or 1s) of information. I have a 64GB flash drive on my key ring; sixty-four gigabytes is sixty-four billion bits—so on this flash drive I could store the information for sixty-four million universes!

Fast-forward to today. In order to describe the universe, it would be necessary to record the location and type of every atom, the energy state of every electron in every atom, and so on. Instead of a mere 1,000 bits, 1 followed by 89 zeroes bits would be needed to describe the universe. So, the question is: if the universe started out so simple, with pretty much no information, where did all the information, all the complexity, come from? Why are there galaxies and stars, atoms and iPhones, rainbows and roses?

A vital clue comes from a simple observation: the reflection of your face in a window. Staring out through the window in your home, maybe you can see cars driving past, trees swaying in the breeze, a dog being walked. Crucially, however, you can also see a faint reflection of your face. The reason is that glass is not perfectly see through. About 95 percent of light goes through and about 5 percent is reflected back.

This fact about the world became extremely difficult to understand at the start of the twentieth century, when physicists discovered that light is a stream of tiny machine-gun bullets called photons—all identical. After all, if they are all identical, surely they should be affected identically by a windowpane? Either they should all go through or they should all be reflected.

There is only one way to explain the fact that 95 percent of photons go through and 5 percent bounce back: that each has a 95 percent chance of being transmitted and a 5 percent chance of being turned back. But this means that, if you could follow an individual photon as it headed towards a windowpane, you could never know for sure whether it would be reflected or transmitted. You could know only the *chance* of it doing one thing, or the *chance* of it doing another. What it actually does is fundamentally unpredictable.

And what is true of photons is true of all denizens of the submicroscopic world: atoms, electrons, neutrinos, everything. At its most basic level, the universe is fundamentally unpredictable, fundamentally random. Arguably, this is the single most shocking discovery in the history of science. In fact, it so upset Albert Einstein that he famously declared: "God does not play dice with the universe." (Less well-known is the retort of quantum physicist Niels Bohr: "Stop telling God where to throw his dice.") However, not only was Einstein wrong; he was spectacularly wrong.

Information is synonymous with randomness. If I have a number that is non-random—say, one repeated a billion times—I can tell you what it is in only a few words: "One repeated a billion times." It therefore contains hardly any information. If, on the other hand, I have a random number that is a billion digits long, then to communicate it to you I must recount every one of the billion digits. It therefore contains a lot of information.

So here is the answer to the conundrum of where the universe's information ultimately comes from. Every random quantum event since the Big Bang has injected information into the universe. Every time an atom spat out a photon—or did not spit out a photon—it injected information; every time an atomic nucleus disintegrated—or did not disintegrate—it injected information.

Not only does Einstein's metaphorical God play dice with the universe, but if He did not there would *be* no universe—certainly not one of the complexity needed for humans to have arisen and for you to be reading these words. We live in a random reality. We live in a universe ultimately generated by the quantum roll of a die.

49

CREDIT CARD COSMOS

Believe it or not, we may be living
in a giant hologram

> "There is a theory which states that if ever for any reason
> anyone discovers what exactly the universe is for and why it is
> here it will instantly disappear and be replaced by something even
> more bizarre and inexplicable. There is another that states
> that this has already happened."
> —DOUGLAS ADAMS[1]

THESE DAYS CREDIT CARDS often feature a hologram—a two-dimensional representation of an object that has every appearance of a three-dimensional one. The idea that the universe is similarly a holographic illusion seems straight out of the pages of a science-fiction novel. Yet there is mounting evidence that this is true. The first hint came, peculiarly, from considerations not of the universe but of black holes.

A black hole is the endpoint of the evolution of a massive star. With its fuel exhausted, the star can no longer generate the internal heat required to push back against the force of gravity trying to crush it. It embarks on a catastrophic runaway shrinkage, its gravity intensifying until nothing, not even light, can escape.

However, in 1974, Hawking discovered something extraordinary and unexpected about black holes: they are not entirely black.

Hawking was thinking about a black hole's horizon, the point of no return for light and matter falling into the hole. And he was thinking about the theory of atoms and their constituents. In quantum theory, the vacuum is not empty. Far from it. It is a roiling sea of *quantum fluctuations*, which can be thought of as subatomic particles and their *antiparticles* popping into existence out of nothing.[2] The law of conservation of energy states that energy can be neither created nor destroyed. However, the quantum twist is that it turns a blind eye as long as each particle-antiparticle pops back out of existence, or annihilates, in a very short time. Such particles, in recognition of their ephemeral status, are dubbed "virtual."

What Hawking realized is that the horizon of a black hole has profound implications for the creation and annihilation of virtual particle-antiparticle pairs. The reason is that one member of such a pair can plummet though the horizon of a black hole while the other careens off into free space. With no partner with which to annihilate, the escaping particle is no longer transient; it is elevated from virtual to real status. Of course, the energy to give it permanent existence must come from somewhere. And Hawking realized that it comes from the gravitational energy of the black hole itself.

The sleet of escaping particles is called *Hawking radiation*, and this causes a black hole to gradually evaporate. Eventually, it disappears altogether.

For a stellar-mass black hole, it would take much longer than the age of the universe to vanish. Nevertheless, Hawking radiation poses a huge problem for physics. Once a black hole—which is no more than a patch of grossly warped space-time anyway—evaporates, there is literally nothing left. The question is: what happens to the information that described the original star that spawned the black hole—information that described, for instance, the location and type of every one of the star's atoms and electrons? It is a fundamental principle in physics that information can neither be created nor destroyed.[3]

A clue comes from the horizon itself. In 1972, the Israeli physicist Jacob Bekenstein discovered that the horizon has a large and inexplicable entropy—a physical quantity associated with microscopic disorder, as previously explained. In 1993, Dutch Nobel Prize-winner Gerard 't Hooft proposed that, contrary to predictions of Einstein's theory of gravity, the horizon of a black hole may not be smooth but rather, on the microscopic scale, fantastically irregular like a miniature mountain landscape. The information describing the original star might therefore be encoded in some way in the horizon's microscopic lumps and bumps. As a black hole evaporates, the irregularities of the horizon would impress themselves on the Hawking radiation in much the same way that music or speech impresses itself on the *carrier wave* of a radio station. If this is the case, then in the evaporation of the black hole no information is actually lost. All is returned to the universe, albeit in a rather garbled form.

The connection with the universe is that, like a black hole, it is surrounded by a horizon. Because the universe was born 13.82 billion years ago, we can see objects whose light has taken less than 13.82 billion years to get to us. This defines the *cosmic light horizon*. Beyond it are objects whose light will take more than 13.82 billion years to reach us. Their light is still on its way to Earth, so we have yet to see them.

Both 't Hooft and the American physicist Leonard Susskind suggested independently that, just as the information to describe a 3D star is encoded in the horizon of a black hole, the information to describe the universe might be encoded in the horizon of the universe. It would make the universe we see around us a hologram—in some sense, a three-dimensional projection of the two-dimensional information on the horizon. You, me, and everything in the universe are holograms.

This may seem very vague and hand-waving. However, in 1998, the Argentinean-American physicist Juan Maldacena published a paper that shored up the idea that we live in a "holographic universe" and set the world of physics alight. Maldacena discovered that a quantum theory that lives on the horizon of the universe can generate a universe inside the boundary that experiences Einstein's theory of gravity. This not only hinted at a long-sought-after connection between quantum theory and Einstein's theory of gravity—which explains why Maldacena's paper is the most cited in the past twenty years—but also elevated 't Hooft and Susskind's speculation that the universe is a hologram to a more evidence-based level.[4]

The only fly in the ointment is that the result Maldacena proved is for a weird space-time called *anti-de Sitter* space. As Einstein recognized, space-time is warped by the presence of matter. But anti-de Sitter space represents a universe whose space-time is warped in a different way to ours. The challenge is now for physicists to show that Maldacena's result also applies in the regular space of our universe and prove beyond any doubt that we are indeed all holograms.

50.

THE UNIVERSE NEXT DOOR

Out there in the universe there are an infinite
number of copies of you reading an infinite
number of copies of this

"There are two things you should remember when dealing
with parallel universes. One, they're not really parallel,
and two, they're not really universes."
—DOUGLAS ADAMS[1]

FAR, FAR AWAY, IN a galaxy remarkably similar to the Milky Way, there is a star that looks remarkably like the sun. And, on the star's third planet, which looks remarkably like the earth, lives someone who, for all the world, looks just like you. They could be your identical twin; not only do they look the same as you, but they are reading this same book too—in fact, they are concentrating on this very line…Actually, it is weirder than this. Far weirder. There are an infinite number of galaxies that look just like our own galaxy containing an infinite number of versions of you, whose lives, up until this moment, have been absolutely identical to yours.

Your doppelgangers live in regions of space beyond the edge of the *observable universe*. If you think their existence is pure science fiction, think again. It is actually an unavoidable consequence of the standard theory of our universe and the standard theory of physics. If you could travel far enough across the universe, you would inevitably run into one of your doubles. In fact, it is even possible to calculate how far you would have to go to meet your nearest doppelganger. The answer is about $10^{10^{28}}$ meters.

To say this is a big number is rather an understatement. In scientific notation, the number 10^{28} is 1 followed by 28 zeroes, which is 10 billion billion billion. Consequently, $10^{10^{28}}$ is 1 followed by 10 billion billion billion zeroes. It corresponds to a distance enormously further than the furthest limits probed by the world's biggest telescope. But do not get hung up on the size of this number. The point is not that your nearest double is at a mind-bogglingly great distance. The point is that you have a double at all.

As already said, this is a consequence of the standard theory of cosmology, which provides a good idea of what lies beyond the edge of the observable universe. But what exactly is the observable universe? And if there is an observable universe, there must be, by implication, an *unobservable* universe. Why can't we see the universe in its entirety?

This is down to two things: the finite speed of light and the fact that the universe was born.[2] Because everything—matter, energy, space, and even time—burst into being in the Big Bang about 13.82 billion years ago, we can see only those objects in the universe whose light has taken less than 13.82 billion years to reach us. Objects whose light would take longer than 13.82 billion years to get here, well, their light is still on its way to us, so we cannot see them yet. Consequently, our telescopes show us only the galaxies—about two trillion of them—within a sphere of space centerd on the earth. This is the observable universe.

As previously discussed, the observable universe is bound by a horizon, which is pretty much like the horizon at sea. Just as we know there is more of the ocean over the horizon, we know there is more of the universe over the universe's *light horizon* (actually, since the universe expanded, or *inflated*, faster-than-light, during its first split-second of existence, the horizon is about forty-two billion light years away).[3] So, what is it like beyond the horizon?

According to the standard picture of the universe, which incorporates the early super-fast epoch of inflation, there is in effect an infinite amount of space beyond the observable universe.[4] Imagine our universe with its two trillion galaxies as inside a sphere like a soap bubble. Well, beyond our soap bubble are an infinite number of other similar-sized soap bubbles. What is it like in the other bubbles?

Each would have had a Big Bang like our own—in fact, a bit of the same one. But, out of the cooling debris of the Big Bang fireball, there would have congealed different galaxies and different stars and different planets. This is because the seeds of large-scale cosmic structures are believed to have been tiny quantum convolutions of the vacuum impressed on it in the first split-second of the universe's existence. Like everything quantum, these were random, both in size and location. And they would have permitted an infinite number of possible universes, each looking different. Or, to put it another way, there would be an infinite number of different cosmic histories, each one played out in its own bubble somewhere in the greater universe.

Actually, it is worse than this.

Quantum theory, our best description of the submicroscopic world, tells us that the universe is ultimately grainy. This means that, if it were possible to cut a volume of space in half, then in half again, and keep on going, then sooner or later, we would come to a tiny volume of space that could not be cut in half any more. Space on ultra-small distances, known as the *Planck scale*, is therefore believed to be like a 3D chessboard. And, just as there are only a finite number of places to put chess pieces on a normal chessboard, at the beginning of the universe there were only a finite number of locations for quantum convulsions to create the seeds of today's galaxy clusters. This means that there are only a finite number—rather than an infinite number—of possible cosmic histories.

If there are only a finite number of histories but an infinite number of places for those histories to be played out, then every history is played out not once but an infinite number of times. Consequently, as mentioned at the beginning of this chapter, there are an infinite number of the places in the universe that contain copies of you who are reading this same book—and, in fact, concentrating on this very line. And there are an infinite number of regions where Donald Trump did not become President of the United States. And an infinite number of places in the universe where the dinosaurs were not wiped out by an asteroid sixty-six million years ago but went on to develop intelligence and build motor cars.

Such endless repetition may be hard to stomach. But cosmologists such as Alex Vilenkin of Tufts University in Massachusetts are philosophical. They say nature has seen fit to take the pattern of a star and repeat it pretty much endlessly and wastefully. Why not also universes?

It is worth emphasizing that all these endless space domains in which all histories are played out are an unavoidable consequence of our standard picture of the universe combined with the standard picture of physics—quantum theory. If one or both of these turns out to be wrong, then there is a "get-out."

Maybe you find the thought of all possible histories being played out disturbing. Personally, I do not. Why? Well, even if this is the dullest, most boring book you've ever had the misfortune to read, I can console myself with the thought that, in an infinite number of other universes, you thought it was the most brilliant thing you have ever come across and bought a copy as a Christmas present for every last one of your friends!

NOTES

PART ONE: BIOLOGICAL THINGS

1. THE COMMON THREAD

1.   A cell is a tiny bag of gloop with the complexity of a small city. It is the "atom of biology." All organisms are assemblages of cells. As far as we know, there is no life at all except cellular life.

2.   In recent years, the idea that DNA is a blueprint for an organism has been recognized as woefully inadequate. Biologists were shocked to discover that human DNA codes for a mere 24,000 genes (genes contain the information to build "proteins": large Swiss Army-knife molecules that carry out a multitude of tasks, such as speeding up chemical reactions, providing scaffolding for cells, and so on). This is not enough to specify a human being. Instead, in a bewilderingly complex way, genes are switched on and off by other genes and by concentrations of chemicals in the environment. It means that the human genome reads differently at different times in the development of an embryo—in effect making it appear like more than one set of 24,000 genes.

3.   Adenine (A), guanine (G), cytosine (C), and thymine (T) are molecules known as "bases." They form the backbone of the "double-helix" of the giant DNA molecule. Each group of three are the bases codes for a different *amino acid*—for instance, TGG codes for Tryptophan. Amino acids are the Lego building blocks of proteins.

4.   *The Medusa and the Snail* by Lewis Thomas (Penguin, 1995).

2. CATCH ME IF YOU CAN

1.   *Dr Tatiana's Sex Advice to all Creation* by Olivia Judson (Vintage, 2003).

2.   The information to build an organism is encoded in deoxyribonucleic acid, or DNA, a double-helical molecule inside each cell. A stretch of DNA that encodes a protein is called a gene. Proteins are the workhorses of the cell, giant molecules assembled from amino-acid building blocks.

3.   "A New Evolutionary Law" by Leigh Van Valen (*Evolutionary Theory*, Vol. 1, p. 1, 1973).

4. *The Red Queen: Sex and the Evolution of Human Nature* by Matt Ridley (Penguin, 1994).
5. "Running with the Red Queen: Host-Parasite Coevolution Selects for Biparental Sex" by Levi Morran et al. (*Science*, Vol. 333, p. 216, 2011).

3. THE OXYGEN TRICK

1. *A Course of Six Lectures on the Chemical History of a Candle* by Michael Faraday (Griffin, Bohn & Co., 1861).
2. The energy liberated by the liquid hydrogen and liquid oxygen fuel is not quite enough to boost into space their combined weight plus that of the metal skin of a rocket. This is why a rocket is built in stages. By dropping off a stage when it has climbed high into the air, a rocket becomes lighter. Its fuel then has an easier job of boosting it into space.
3. The electrons in an atom are arranged in "shells," each with a maximum complement of electrons. Having a complete shell is hugely desirable for an atom. Hydrogen can achieve this by losing an electron (in fact, its only electron); oxygen by gaining two electrons. This is why an oxygen atom grabs electrons from two hydrogen atoms. The state in which two hydrogen atoms lose an electron and an oxygen atom gains two electrons is the lowest energy, desirable state, the equivalent of a ball lying at the foot of the hill.
4. A proton, which is roughly 2,000 times more massive than an electron, is one of the two constituents of the core, or nucleus, of an atom. The other is a neutron. All atomic nuclei contain both particles, apart from the nucleus of a hydrogen atom, which contains only a proton.
5. Naively, it might be thought that an electron simply slams into a proton, driving it through a pore in the cell membrane. Actually, the electron changes the shape of a protein—it has one shape without the electron and another with the electron. Such shape changes are what forces a proton across the membrane.

4. SEVEN-YEAR ITCH

1. *Sweet Dreams: Philosophical Obstacles to a Science of Consciousness* by Daniel Dennett (MIT Press, 2006).
2. *The Lives of a Cell* by Lewis Thomas (Penguin, 1978).
3. *Cosmos: A Personal Voyage* written by Carl Sagan, Ann Druyan, and Steven Soter (Public Broadcasting Service, 1980).
4. "The Secrets of the Human Cell" by Peter Gwynne, Sharon Begley, and Mary Hager (*Newsweek*, 20 August 1979, p. 48).
5. DNA, or deoxyribonucleic acid, is the giant biomolecule that stores, in encoded form, the structure of a cell's proteins.

5. LIVING WITH THE ALIEN

1. "Revised Estimates for the Number of Human and Bacteria Cells in the Body" by Ron Sender, Shai Fuchs, and Ron Milo (*PLOS Biology*, 19 August 2016: https://doi.org/10.1371/journal.pbio.1002533).
2. NIH Human Microbiome Project (https://hmpdacc.org).
3. "Gut Microbiome Alterations in Alzheimer's Disease" by Nicholas Vogt et al. (*Nature*, 19 October 2017: https://www.nature.com/articles/s41598-017-13601-y).

6. THE DISPENSABLE BRAIN

1. "The juvenile sea squirt wanders through the sea searching for a suitable rock or hunk of coral to cling to and make its home for life. For this task, it has a rudimentary nervous system. When it finds its spot, and takes root, it doesn't need its brain anymore, so it eats it. It's rather like getting tenure." From *Consciousness Explained* by Daniel Dennett (Penguin, 1993).
2. "Meet the Creature that Eats Its Own Brain!" by Steven Goodhart (https://goodheartextremescience.wordpress.com/2010/01/27/meet-the-creature-that-eats-its-own-brain/).
3. "The Electric Brain," (*NOVA*, 23 October 2001: www.pbs.org).
4. *Society of the Mind* by Marvin Minsky (Pocket Books, 1988).
5. *In the Palaces of Memory: How We Build the Worlds Inside Our Heads* by George Johnson (Vintage, 1992).
6. "Are Brains Shrinking to Make Us Smarter?" by Jean-Louis Santini (6 February 2011: https://phys.org/news/2011-02-brains-smarter.html).
7. "How Humans (Maybe) Domesticated Themselves" by Erika Engelhaupt (*Science News*, 6 July 2017: https://www.sciencenews.org/article/how-humans-maybe-domesticated-themselves).
8. As quoted by Emerson Pugh's son in *The Biological Origin of Human Values* by George E. Pugh (Routledge & Kegan Paul, 1978).

PART TWO: HUMAN THINGS

7. INTERACTION, INTERACTION, INTERACTION

1. Hominin is the term used to refer to the group that includes modern humans, extinct human species, and all our immediate ancestors (including members of the genera Homo, Australopithecus, Paranthropus, and Ardipithecus).
2. The dependence of large communities on crops also made them vulnerable to famines caused by failures of those crops. And when many people lived together in close proximity, diseases spread easily—sometimes with devastating effects.

8. THE GRANDMOTHER ADVANTAGE

1. "The Origin of Menopause: Why Do Women Outlive Fertility?" by Tabitha Powledge (*Scientific American*, 3 April 2008).

9. LOST TRIBE

1. *Guns, Germs, and Steel: A Short History of Everybody for the Last 13,000 Years* by Jared Diamond (Vintage, 1998).
2. For more on this, see *What a Wonderful World* by Marcus Chown (Faber, 2014).
3. "Oldest Known Human Fossil Outside Africa Discovered in Israel" by Hannah Devlin (*Guardian*, 25 January 2018: https://www.theguardian.com/science/2018/jan/25/oldest-known-human-fossil-outside-africa-dis-covered-in-israel).
4. "Neandertal DNA" by Mark Rose (*Archaeology*, Vol. 50, Number 5, September/October 1997).
5. "Neanderthal Artists Made Oldest-Known Cave Paintings" by Emma Marris (*Nature*, 23 February 2018: https://www.nature.com/articles/d41586-018-02357-8).
6. "Palaeoanthropology: The Time of the Last Neanderthals" by William Davies (*Nature*, Vol. 512, p. 260–261, 21 August 2014: https://www.nature.com/articles/512260a).
7. "A Draft Sequence of the Neandertal Genome" by Richard Green et al (*Science*, Vol. 328, p. 710, May 2010: http://science.sciencemag.org/content/328/5979/710.full).

10. MISSED OPPORTUNITY

1. "Neil Armstrong's Photo Legacy: Rare Views of First Man on the Moon" by Robert Pearlman (Space.com, 27 August 2012: https://www.space.com/17308-neil-armstrong-photo-legacy-rare-views.html).
2. "Man on the Moon—Neil Armstrong's Iconic Photograph" (*Amateur Photographer*, 24 August 2017: http://www.amateurphotographer.co.uk/iconic-images/moon-iconic-photograph-neil-armstrong-18051).
3. "The Apollo Astronaut Who Was Allergic to the Moon" by Lucas Reilly (*Mental Floss*, 6 February 2017: http://mentalfloss.com/article/91628/apollo-astronaut-who-was-allergic-moon).
4. "The Footprints at Laetoli" by Neville Agnew and Martha Demas (http://www.getty.edu/conservation/publications_resources/newsletters/10_1/laetoli.html).

NOTES

PART THREE: TERRESTIAL THINGS

11. THE ALPHABET OF NATURE

1. *The Feynman Lectures on Physics*, Vol. II, p. 1–10 (Addison-Wesley, New York, 1989).
2. By this time, it had been realized that atoms are made of even smaller things—electrons, which are believed to be elemental, and protons and neutrons, which are made from quarks—but the fundamental idea that atoms are nature's basic Lego bricks remains true.

13. DEEP IMPACT

1. An asteroid is a small rocky body that orbits the sun. Large numbers of them are found between the orbits of Mars and Jupiter. The largest is Ceres, which is 946 kilometers across and was discovered on January 1, 1801. Collisions between asteroids or the effect of Jupiter's powerful gravity can cause an asteroid to be ejected from the "main belt." If its orbit crosses that of the earth, it poses a serious threat to our planet.
2. "Chelyabinsk Meteor: Wake-Up Call for Earth" by Elizabeth Howell (Space. com, 2 August 2016: https://www.space.com/33623-chelyabinsk-meteor-wake-up-call-for-earth.html).
3. "Extraterrestrial Cause for the Cretaceous-Tertiary Extinction" by Luis Alvarez et al (*Science*, Vol. 208, p. 1095, 6 Jun 1980: http://science.sciencemag.org/content/208/4448/1095).
4. "Why Some Species Thrived When Dinos Died" by Sid Perkins (*Science*, 24 July 2013: http://www.sciencemag.org/news/2013/07/why-some-species-thrived-when-dinos-died).
5. "Site of Asteroid Impact Changed the History of Life on Earth: the Low probability of Mass Extinction" by Kunio Kaiho and Naga Oshima (*Nature Scientific Reports*, Vol. 7, No. 14855, 9 November 2017: https://www.nature.com/articles/s41598-017-14199-x).

14. SECRET OF SUNLIGHT

1. Actually, a small amount of heat is trapped in the atmosphere by the greenhouse gas carbon dioxide, the by-product of the burning of fossil fuels, causing the planet to gradually warm up.
2. Absolute zero is the lowest temperature attainable. When an object is cooled, its atoms move more and more sluggishly. Absolute zero, which is equivalent to -273.15 in the Celsius scale, is the temperature at which atoms stop moving altogether (though actually this is not entirely true, since the Heisenberg Uncertainty Principle produces a residual quantum jitter even at absolute zero).

3. *Four Laws that Drive the Universe* by Peter Atkins (Oxford University Press, 2007).

PART FOUR: SOLAR SYSTEM THINGS

16. KILLER SUN

1. *The Three* by Sarah Lotz (Hodder, 2015).
2. The impact of the solar flare distorted the earth's magnetic shield, so solar particles were no longer merely funnelled down at the poles but could reach anywhere on the planet. The collision of solar particles with atoms in the atmosphere caused those atoms to glow and create the aurora.
3. *The Sun Kings: The Unexpected Tragedy of Richard Carrington and the Tale of How Modern Astronomy Began* by Stuart Clark (Princeton University Press, 2009).
4. "A Scary 13th: 20 Years Ago, Earth was Blasted with a Massive Plume of Solar Plasma" by Adam Hadhazy (*Scientific American*, 13 March 2009: https://www.scientificamerican.com/article/geomagnetic-storm-march-13-1989-extreme-space-weather/).

17. LIGHT OF OTHER DAYS

1. Actually, every time a photon is deflected, or scattered, by a free electron in the sun, it is sapped of a little energy. So, although the nuclear reactions at the heart of the sun create high-energy gamma-ray photons, by the time they emerge from the solar surface, or photosphere, they are low-energy photons of visible light (the lost energy is of course what keeps the sun hot). Thus, strictly speaking, although sunlight takes 30,000 years to emerge from the sun, the photons of which it is composed are not the same photons that started out on the journey so many centuries before.

18. A BRIEF HISTORY OF FALLING

1. *Life, the Universe and Everything* by Douglas Adams (Picador, 2002).
2. Actually, the moon is orbiting the earth in an elliptical path, but, to a good approximation, it is a circle.
3. In fact, this was the key insight that led Einstein to his theory of gravity—the general theory of relativity—in 1915.
4. *The Principia's* full title was *Philosophiæ Naturalis Principia Mathematica* (*The Mathematical Principles of Natural Philosophy*). It was published in two volumes on July 5, 1687.

19. THE PLANET THAT STALKED THE EARTH

1. A Lagrange point is one of the five locations in the sun-Earth system where the gravitational and centrifugal forces on a body balance so that, in principle, it can stay becalmed forever.
2. See Chapter 12: "Rock Sponge."

20. PLEASE SQUEEZE ME

1. See Chapter 12: "Rock Sponge."

21. HEX APPEAL

1. "A Laboratory Model of Saturn's North-Polar Hexagon" by Barbosa Aguiar et al. (*Icarus*, Vol. 206, p. 755, 2010).
2. "Meandering Shallow Atmospheric Jet as a Model of Saturn's North-Polar Hexagon" by Raúl Morales-Juberías et al. (*Astrophysical Journal Letters*, Vol. 806, Number 1, 10 June 2015).

22. MAP OF THE INVISIBLE

1. Actually, he called it "George's Star."
2. The existence of Neptune was predicted independently by Englishman John Couch Adams. When they met, Le Verrier and Adams became firm friends, and today the discovery of Neptune is usually attributed to both Adams and Le Verrier.
3. See *The Hunt For Vulcan: How Albert Einstein Destroyed a Planet and Deciphered the Universe* by Thomas Levenson (Head of Zeus, 2016).

23. LORD OF THE RINGS

1. "Density waves in Saturn's rings" by Fraser Cain (*Universe Today*, 10 November 2004: https://www.universetoday.com/10034/density-waves-in-saturns-rings/).

24. STARGATE MOON

1. The final words of Dave Bowman as he enters the monolith in *2001: A Space Odyssey* by Arthur C. Clarke (Orbit, 2001).
2. "Solving the Mystery of Iapetus" by Paulo Freire (*Geophysical Research Letters*, Vol. 33, p. L16203: https://arxiv.org/pdf/astro-ph/0504653.pdf).
3. "Delayed Formation of the Equatorial Ridge on Iapetus from a Subsatellite Created in a Giant Impact" by Andrew Dombard et al (*Journal of Geophysical Research—Planets*, Vol. 117, Issue E3, March 2012).

4. "How Saturn's Icy Moons Get a (Geologic) Life" by Richard Kerr (*Science*, Vol. 311, p. 29, 6 Jan 2006).

PART FIVE: FUNDAMENTAL THINGS

25. INFINITY IN THE PALM OF YOUR HAND

1. Actually, a quantum wave associated with a particle is a weird type of wave. It is an abstract, mathematical entity that is imagined to fill all of space. Where the wave is big—strictly speaking, where it has a large amplitude— there is a high chance, or probability, of finding the particle, and where the wave is small, a low probability of finding it.
2. The reason for this is that a low-energy/low-mass particle like the electron naturally has a low-energy quantum wave associated with it. Imagine a low-energy wave on the surface of a pond. It is sluggish and its characteristic size—the distance between successive wave peaks—is large.
3. Actually, in an unexpected twist, quantum theory tells us that empty space is not completely empty. It is a roiling sea of the "zero-point fluctuations of the quantum fields." But that's another story!
4. This was the biggest discrepancy between a prediction and an observation in the history of science until 1998. That year marked the discovery of the dark energy that fills all of space and whose repulsive gravity is speeding up the expansion of the universe. When quantum theory is used to predict the energy of the vacuum—the dark energy—it yields a number that is 1 followed by 120 zeroes bigger than observed. This is a strong indication that our current theory of physics is inadequate!

26. BUNGALOW BENEFITS

1. According to Einstein's special theory of relativity of 1905, time flows more slowly for someone moving relative to another person. And, since the top of a building is moving faster than the bottom simply by virtue of the rotation of the earth, this effect counteracts the effect of gravity slowing time. However, it is relatively small and does not alter the conclusion that you age faster on the top floor of a building than on the ground floor.
2. "Optical Clocks and Relativity" by James Chin-Wen Chou et al. (*Science*, Vol. 329, p. 1,630, 24 September 2010).
3. "String Theory: From Newton to Einstein and Beyond" by David Berman (https://plus.maths.org/content/string-theory-newton-einstein-and-beyond).

27. THE INCREDIBLE EXPLODING MOSQUITO

1. *The Feynman Lectures on Physics*, Vol. II, p. 1–10 (Addison-Wesley, New York, 1989).
2. To be precise, the electromagnetic force between a proton and an electron in a hydrogen atom is 10^{40} times stronger than the gravitational force between them. Hydrogen, the lightest atom in nature, consists of an electron orbiting a nucleus containing a single proton.

28. THE UNKNOWABLE

1. For more on incomputability, see Chapter 6: "God's Number," in *The Never-Ending Days of Being Dead* by Marcus Chown (Faber, 2007).

29. DOUBLE TROUBLE

1. To be precise, the probability of finding a particle at a particular location in space—a number between 0 and 1, with 0 corresponding to a 0 percent chance and 1 to a 100 percent chance—is the square of the height of the quantum wave at that location (the amplitude is actually a complex number, but that is another story!).
2. It is always possible to arrange a point of view—or frame of reference—from which the ricocheting bowling balls fly in opposite directions.

30. LOOPY LIQUID

1. Absolute zero is the lowest temperature attainable. On the Celsius scale it corresponds to -273.15 °C and on the Kelvin scale to 0 Kelvin.
2. See Chapter 29: "Double Trouble."

32. WHO ORDERED THAT?

1. *Rendezvous with Rama* by Arthur C. Clarke (Gollancz, 2006).
2. See Chapter 17: "Light of Other Days."
3. The law of conservation of energy says that energy cannot be created or destroyed, only transformed from one type of energy into another. As Einstein showed in 1905, mass is simply another form energy. The energy of the quark elastic—strictly speaking, the gluon fields—can therefore be converted into the mass-energy of new quarks.
4. See Chapter 39: "Stardust Made Flesh."
5. There could be more generations of neutrinos if those neutrinos are of a type known as *sterile*. The normal neutrinos, although antisocial, do interact with normal matter very occasionally via nature's *weak nuclear force*.

Sterile neutrinos would not even do this. Their sole interaction with normal matter would be via the gravitational force, making them nigh on impossible to detect directly.

33. A WONDERFUL THING IS A PIECE OF STRING

1. The three space dimensions are east-west, north-south, and up-down. And, of course, there is the time dimension: past-future. The twist is that Einstein's special theory of relativity of 1905 shows that space and time are actually aspects of the same thing, something apparent only to an observer traveling close to the speed of light. In recognition of this fact, physicists talk of space-time, a seamless amalgam of four space-time dimensions.
2. The exchange particles that give rise to nature's fundamental forces are not quite like the kind with which we are familiar. They are known as virtual particles. See *QED: The Strange Theory of Light and Matter* by Richard Feynman (Penguin, 1990).

34. NO TIME LIKE THE PRESENT

1. Letter written by Albert Einstein to the bereaved family of Michele Besso when his long-standing friend died in 1955.
2. "On the Electrodynamics of Moving Bodies" by Albert Einstein (*Annalen der Physik*, Vol. 17, p. 891, 1905).

35. HOW TO BUILD A TIME MACHINE

1. See Chapter 26: "Bungalow Benefits."
2. A black hole is a region of space where gravity is so strong that nothing can escape—not even light, hence its blackness.

PART SIX: EXTRATERRESTIAL THINGS

36. OCEAN WORLDS

1. *2010: Odyssey Two* by Arthur C. Clarke (Harper Collins, 2000).
2. See Chapter 20: "Please Squeeze Me."
3. "Life on the Ocean Floor, 1977" by Cristina Luiggi (*The Scientist*, 1 September 2012: https://www.the-scientist.com/?articles.view/articleNo/32523/title/Life-on-the-Ocean-Floor--1977/).
4. "Enceladus Shoots Supersonic Jets of Water" by Ashley Yeager (Nature.com, 26 November 2008: http://www.nature.com/news/2008/081126/full/news.2008.1254.html).

37. ALIEN GARBAGE

1. "On the Possibility of Extraterrestrial Artefact Finds on the Earth" by A.V. Arkhipov (*The Observatory*, Vol. 116, p. 175, 1996).

38. INTERPLANETARY STOWAWAYS

1. The loss of its internal heat caused the interior of Mars to solidify. And, since it is the circulation of molten material, carrying giant electrical currents, that generates a planetary magnetic field like the earth's, Mars lost its precious magnetic shield.

39. STARDUST MADE FLESH

1. "Song of Myself" by Walt Whitman.
2. The nuclear force, also known as the strong force, has a very short range, which is why its existence was discovered only in the twentieth century when physicists began to probe the nuclei of atoms.
3. The nuclear reactions that create iron, rather than liberating energy (which ultimately emerges from a star as starlight), actually suck energy vampire-like from the star. This makes a star unstable, which is why iron marks the endpoint of element-building inside stars.

40. THE FRAGILE BLUE DOT

1. Carl Sagan's Famous "Pale-Blue Dot," quote, *Cosmos: A Space-time Odyssey* (https://www.youtube.com/watch?v=Cm6NS6uDqt8).

PART SEVEN: COSMIC THINGS

41. THE DAY WITHOUT A YESTERDAY

1. See Chapter 42: "Afterglow of Creation."
2. Strictly speaking, the universe must be either expanding or contracting. We live in a "restless universe" that simply cannot stand still.
3. See *Afterglow of Creation* by Marcus Chown (Faber, 2010) and *The Very First Light* by John Boslough and John Mather (Basic Books, 2008).

42. GHOST COSMOS

1. *The Hitchhiker's Guide to the Galaxy* by Douglas Adams (Pan, 2009).
2. See Chapter 47: "The Voice of Space."

43. HEART OF DARKNESS

1. "Astronomers Say They've Found Many of the Universe's Missing Atoms" by Adam Mann (*Science*, 10 October 2017: http://www.sciencemag.org/news/2017/10/astronomers-say-they-ve-found-many-universe-s-missing-atoms).

2. "Opposite Thermodynamic Arrows of Time" by Lawrence Schulman (*Physical Review Letters*, Vol. 83, p. 5419, 27 December 1999: https://arxiv.org/pdf/cond-mat/9911101.pdf).

3. "Confrontation of MOND Predictions with WMAP First Year Data" by Stacy McGaugh (29 September 2004: https://arxiv.org/abs/astro-ph/0312570v4).

44. AFTERGLOW OF CREATION

1. *The Sentinel* is the 1948 story that was the basis of the movie *2001: A Space Odyssey* by Stanley Kubrick and Arthur C. Clarke (*The Sentinel*, Harper-Voyager, 2000).

2. Actually, although the leftover heat from the Big Bang was originally detected as microwaves—radio waves with the wavelength of a few centimeters—it is most intense at wavelength of a few millimeters. The wavelength of light—more technically known as electromagnetic radiation—is a measure of distance between successive wave crests.

3. Temperature is a measure of microscopic motion. As a body is cooled, its atoms jiggle about ever more sluggishly. Eventually, they stop moving entirely. The temperature at which this happens—the lowest temperature possible—is known as absolute zero.

4. Bell Labs was part of the giant AT&T phone company, which employed more than a million people at multiple sites across the United States.

45. MASTERS OF THE UNIVERSE

1. A light year is the distance that light, which in a vacuum moves at 299,792 kilometers per second, travels in a year. It is about 9.5 trillion kilometers.

2. *Gravity's Engines: How Bubble-Blowing Black Holes Rule Galaxies, Stars, and Life in the Cosmos* by Caleb Scharf (Scientific American Books, New York, 2012).

46. FLIPPING GRAVITY

1. Strictly speaking, energy-momentum, which is known as a 4-vector.

2. Actually, in Einstein's theory of gravity, the source of gravity is (energy density + 3 X pressure).

47. THE VOICE OF SPACE

1. In the explosion of even the biggest H-bomb only about a kilogram or so of mass disappears, converted into the heat of the nuclear fireball.

2. Since gravitational waves were first detected on Earth, they have been picked up from a total of five events. Four were from the merger of pairs of black holes and one from the merger of super-compact neutron stars. The latter event, observed on August 17, 2017, is the most significant because in addition to gravitational waves it created light, which was picked up by telescopes all over the world. Analysis of the light revealed that the fireball forged at least ten times the mass of the earth in pure gold. Scientists have long wondered where gold came from. Now, at last, they know.

3. Sadly, the prize was not shared by Scotsman Ron Drever. I was a physics graduate student at the California Institute of Technology in Pasadena, one of the two institutions behind LIGO, when the prototype was being built, and I remember going to a talk by Drever. I remember him carrying his papers in two supermarket carrier bags and his overhead-projector transparencies were covered in tea stains and fingerprints. Drever, a key member of the LIGO team, was an experimental genius (he actually built a TV from scratch on which his family watched the Coronation of the Queen in 1953). Unfortunately, the Scottish physicist was apparently not good at sharing control of the project, and he was fired in 1995. He lived on in Pasadena. He did not marry and had few friends. Eventually, he began to suffer from dementia. In the end, Caltech professor Peter Goldreich flew him to New York and put him on a plane back to Glasgow to be met by his brother. From there he went into a care home in Scotland. Unfortunately, he died on March 7, 2017, seven months before the award of the Nobel Prize for the discovery of gravitational waves.

49. CREDIT CARD COSMOS

1. *The Restaurant at the End of the Universe* by Douglas Adams (Pan, 2009).

2. Every subatomic particle has an associated antiparticle with opposite properties such as electrical charge. For instance, the negatively charged electron is twinned with a positively charged antiparticle known as the positron. When a particle is created from the vacuum it is always together with its antiparticle. When a particle and its antiparticle meet, they self-destruct, or annihilate, in a flash of high-energy light, or gamma rays.

3. The reason for this is that the laws of physics describe the future in terms of the present. For instance, the location of the moon tomorrow is determined from its location today by applying Newton's universal law of gravity. The past is contained in the future. No information is lost.

4. "The Large N Limit of Superconformal Field Theories and Supergravity" by Juan Maldacena (*Advances in Theoretical and Mathematical Physics*, Vol. 2, p. 231, 1998: http://arxiv.org/pdf/hep-th/9711200.pdf).

50. THE UNIVERSE NEXT DOOR

1. *The Hitchhiker's Guide to the Galaxy* by Douglas Adams (Pan, 2016).
2. See Chapter 41: "The Day Without a Yesterday."
3. Although it is true that material object can travel faster than light—or even at the speed of light—Einstein's general theory of relativity of 1915 permits space to expand at any rate it likes.
4. Inflation has been likened to the explosion of an H-bomb compared with the stick of dynamite of the Big Bang that took over when inflation ran out of steam after the first split-second of the universe's existence. It is believed to have been driven by the quantum vacuum. Not the mundane vacuum we see around today but a high-energy vacuum with the unusual property of repulsive gravity. Despite inflation's success in explaining some puzzling features of our universe, however, the microscopic physics which under-pins the theory is not understood.

ACKNOWLEDGEMENTS

My thanks to the following people who helped me directly, inspired me or simply encouraged me during the writing of this book. Karen, Jo Stansall, Felicity Bryan, Michele Topham, Manjit Kumar, Dave Hough, Monica Hope, and Patricia Chilver.

ABOUT THE AUTHOR

MARCUS CHOWN is an award-winning writer and broadcaster. Formerly a radio astronomer at the California Institute of Technology in Pasadena, he is cosmology consultant of *New Scientist*. His books include *The Ascent of Gravity* (name the *Sunday Times* 2017 Science Book the Year), *What A Wonderful World*, *Quantum Theory Cannot Hurt You*, and *We Need to Talk About Kelvin* (shortlisted for the 2010 Royal Society Book Prize). Chown has also tried his hand at apps and won the Bookseller Digital Innovation of the Year Award for Solar System for iPad.